Principles of
structural analysis

By the same author

Analysis of statically indeterminate frameworks

Principles of
structural analysis

T. M. Charlton, M.A., F.I.C.E., M.I.C.E.I.

Professor of Civil Engineering in the
Queen's University of Belfast

Foreign Member of the Finnish Academy
of Technical Sciences

Longmans

LONGMANS, GREEN AND CO LTD
London and Harlow
*Associated companies, branches and representatives
throughout the world*

© Longmans Green and Co., Ltd., 1969

First published 1969

SBN   582 44909 X

*Made and printed in Great Britain by
William Clowes and Sons, Limited, London and Beccles*

Contents

Contents

Preface

The purpose of this book is to provide a logical, elementary course in the analysis, in respect of forces and deformation caused by loading, of statically indeterminate structures including frameworks, grids, suspension bridges and arches. It is intended primarily for undergraduate students working for a university degree or its equivalent in civil or structural engineering but the author hopes that practising engineers will find its contents beneficial also.

A thorough grounding in the principles of statics and analysis of statically determinate systems is assumed. Attention is concentrated on principles and their application in terms of elementary mathematical techniques. Thus the two basic approaches to the analysis of statically indeterminate systems are emphasised and developed for linearly elastic structures in terms of flexibility and stiffness coefficients respectively. Energy principles are included for the purpose of facilitating calculations of the flexibility coefficients of linear systems and deflections due to linear and non-linear effects (the principles of virtual work and complementary energy) and for the approximate analysis of both simple and complex systems (the principles of potential and complementary energy, respectively).

Devices of particular value for linear systems such as symmetry and antisymmetry, the reciprocal theorem and model analysis are developed to the extent, it is hoped, of enabling the reader to obtain a thorough appreciation of their potentialities.

The treatment of deflection of beams in Chapter 4 is intended to provide the greatest possible insight into flexural behaviour and in this respect the author gratefully acknowledges the influence of the writings of the late Sir Charles Inglis, F.R.S.

The use of matrices is avoided because it is believed that principles and useful devices of structural analysis tend to be obscured for the beginner thereby, especially since he is then faced with the mastery of two instead of one set of novel principles. Moreover, by matrices, attention is confined usually to the formal analysis of linear systems only. Matrices are, however, introduced in Chapter 9 with a brief explanation of their nature and uses since it is intended that this book be regarded as introductory to the wealth of books on so-called matrix methods of structural analysis which is available nowadays. Thus the presentation of the subject matter in respect of linear behaviour throughout the book is intended to be in accordance with the thought underlying matrix methods.

Informal, numerical methods such as moment distribution are omitted as being inappropriate to this book which is concerned with encouraging concentration on principles and their uses. The essential features of relaxation methods which represent the principle underlying moment distribution and the iterative solution of linear simultaneous equations, are given in Chapter 9.

With the present-day preoccupation with matrix techniques and use of automatic computers there is, perhaps, a tendency to overlook powerful manual devices of approximate analysis by energy principles. It is hoped that Chapter 8 will serve to correct any tendency of that kind.

Since it is believed that an appreciation of the historical development of a subject adds greatly to understanding the subject in the deepest sense, Appendix I is included. Appendix II on the elementary theory of minimum weight is for the purpose of making the student aware of the existence of this work, related as it is to basic principles. It is, perhaps, mainly of academic interest, however, owing to the complexity of the factors which are involved in the design of economical, safe structures.

Numerical examples throughout the book are in metric units using the Newton (N) as the unit of force (10 kN = 1 ton force, very nearly). Dead loads are specified in kg (1 kN = 100 kg force, very nearly) and wind load intensities are specified in kN in accordance with current recommendations.

Finally, the author wishes to express his deep gratitude to Professor Arvo Ylinen of Helsinki and the publishers for their encouragement and to his colleague Mrs. M. Svehla, Dipl. Ing. (Budapest) for assistance and criticism throughout preparation of the work. Also to his secretary Mrs. J. Wilson for typing the manuscript and to Mr. J. Bell for assistance in preparing the illustrations.

Belfast 1968   T. M. CHARLTON

1

Introduction

1.1 Structural analysis is concerned with the calculation of the forces in the members and deformation of structures due to specified loading in relation to the essential requirements of safety, function and economy. By definition, a statically determinate structure can be analysed with respect to the forces in its members by means of the principles of statics alone. The analysis of a statically indeterminate structure, the type of structure with which this book is concerned, necessitates, however, use of the law of elasticity and conditions of compatibility of the strains of its members in addition to the principles of statics. This is because such a structure possesses members, connections and/or supports which are redundant or supernumary to the requirements of statics so that the equations provided by the latter are insufficient. The law of elasticity and conditions of compatibility of the strains of the members provide the means of obtaining the necessary additional equations. If, however, the members of a statically indeterminate structure were made of inelastic rigid material then precise analysis would be impossible. Fortunately, materials of engineering construction deform under load in a predictable manner and so this difficulty does not arise in practice.

The common practice of designing structures to be statically indeterminate arises from considerations of economy having regard to available materials such as uniform rolled-steel joists, self-weight aspects and joint and support details. Thus, a simple structure consisting of a uniform rolled steel beam may be of smaller section to support a specified loading if it is encastré rather than simply supported since, in the latter instance, a large proportion of its material is subjected to only small stresses. Again, the design of a long span bridge might necessitate intermediate supports owing to considerations of the self-weight of the structure and though each span could then be simply supported in its own right, continuity of the main girders usually enables saving in cost to be achieved because of the reduction of the section of the girders which is then possible. The design of statically determinate pin-jointed trusses is also defeated by the costly manufacture and maintenance of pin-joints, rigid joints being preferable with their implication of statical indeterminacy. Again, the erection costs of a structure may be reduced by the introduction of redundants

1*

which enable site work to be expedited; for structural design involves not only the behaviour of the completed structure, but safe erection procedure also.

While analysis enables the effects of specified loading on a statically indeterminate structure to be calculated, this kind of structure suffers from the disadvantage that forces in members can arise from differential thermal expansion, inaccurate fitting and connecting of members and (when the support system is statically indeterminate) differential settlement of supports. These effects are usually covered by the term 'self-straining'. Self-straining forces are often difficult to predict because of lack of data but when the structure is made of ductile material such as mild steel they are usually not detrimental to its ultimate strength. Having regard to self-straining, perhaps the most practically meaningful result of analysis is the deflections caused by specified loading.

1.2 Statical indeterminacy: redundants and degrees of freedom [1]

In order to consider the question of statical indeterminacy, redundants and degrees of freedom, it is important first to specify assumptions, upon which structural analysis is usually based, as follows:

(*a*) structures are of such geometrical form and material that there is no significant change of geometry following the application of working loads (i.e. deflections are small);

(*b*) the members of structures deform in proportion to the forces which they are called upon to exert (such deformation usually being in accordance with Hooke's law of linear elasticity);

(*c*) structures are in stable equilibrium under working loads.

Single-beam systems are, perhaps, the most simple to examine in respect of statical determinacy for they may be regarded as rigid bodies constrained by a system of supports. Treating such systems in one plane only so that there are three degrees of freedom (two degrees of freedom of translation; one degree of freedom of rotation) three components of support is the minimum necessary to provide complete constraint. Thus a single span beam requires supports vertically at each end together with horizontal support or constraint at one end (Fig. 1.1(a)) if it is to carry loads in its plane generally. If the loading is vertical then the horizontal supporting force is zero but loading will generally have a horizontal component also. Clearly, the forces exerted by the supports for any loading of the beam may be calculated by the three equations (namely, of equilibrium vertically, horizontally and in respect of couples) which the principles of statics provide for plane systems.

It is, however, possible to provide three supporting effects which do not provide complete restraint as shown in Fig. 1.1(b). In this instance there is one support which is supernumerary to the requirements of statics if vertical loads only are applied to the system, but the system is not a complete structure

[1] See also reference 1 of the bibliography.

in that a horizontal component of load would cause motion. Thus a fourth (horizontal) support must be provided as shown in Fig. 1.1(c). One of the vertical supports is, in fact, redundant in both instances.

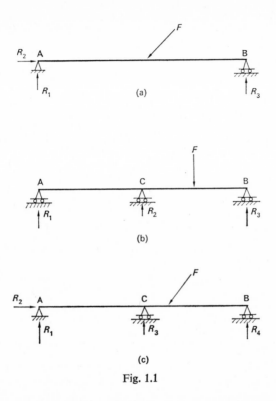

Fig. 1.1

The cantilever (Fig. 1.2(a)) is another statically determinate single-beam system. By virtue of its being built-in or encastré at its root three supporting forces or reactions are possible, namely a vertical force, a horizontal force and a couple. If, however, the cantilever is provided with a support at its free end (Fig. 1.2(b)) it becomes statically indeterminate with four supporting effects, one of which is redundant. In this instance the redundant may be regarded as either the support at the free end or the restraint in respect of a couple at the root.

For three-dimensional or space systems the least number of components of support for complete restraint is six because such systems possess six degrees of freedom as rigid bodies. Components of support in excess of this number are redundant provided that the supports are arranged such that no degree of freedom is restrained in more than one manner.

A framework is said to be statically indeterminate if conditions of equilibrium of its joints are insufficient for the purpose of its analysis. Thus, for

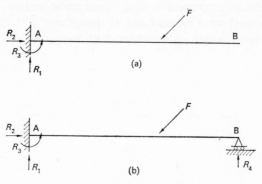

Fig. 1.2

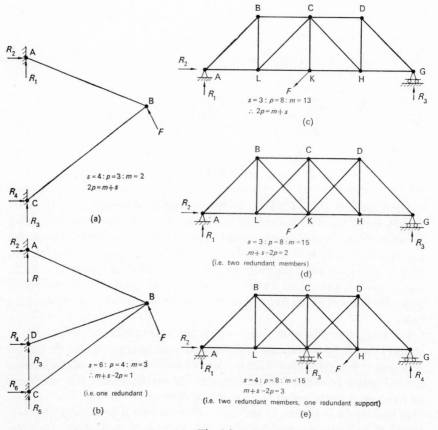

Fig. 1.3

plane frameworks with pin-joints there are two degrees of freedom at each unrestrained joint (that is, two quantities are necessary to define the movement or deflection of a 'free' joint) and two conditions of equilibrium per joint (see Fig. 1.3) so that if there are p joints, m members and s independent components of support, a test of statical determinancy is:

$$2p = m + s \tag{1.1}$$

This is because the left-hand side represents the total possible number of equations of equilibrium for a plane pin-jointed framework while the right-hand side represents the total number of unknown quantities (the forces in the members and in the supporting elements). It follows that if three degrees of freedom are eliminated by support restraints, m is equal to the number of degrees of freedom for a plane statically determinate truss. If

$$2p > m + s \tag{1.2}$$

the system is a mechanism, while if

$$2p < m + s \tag{1.3}$$

the system can be statically indeterminate as a whole with $(m + s - 2p)$ members or supporting elements which are redundant and which could be removed without affecting the function of the framework. Equations (1.1), (1.2) and (1.3) represent necessary but not sufficient conditions. Thus, some additional conditions must be fulfilled as follows:

(*a*) the framework must be made in such a way that the members are connected together to form triangular shapes throughout;
(*b*) the number of degrees of support at any joint must be not greater than two and, since a plane framework as a whole possesses three degrees of freedom, s cannot be less than three if the system is to be a structure.

Similarly, a pin-jointed space framework is statically determinate if

$$3p = m + s \tag{1.4}$$

and is statically indeterminate if

$$3p < m + s \tag{1.5}$$

subject to condition (*a*) above, that the maximum number of components of support at any joint is three and s is not less than six.

Some examples of the use of these conditions for pin-jointed frameworks are given in Fig. 1.3.

The members of rigidly jointed plane frameworks, on the other hand, are subjected to shearing force and bending moment as well as axial force so that there can be as many as three unknowns for each member. Moreover, a couple as well as two forces can operate at a joint: there are three degrees of

5

freedom per 'free' joint. Therefore, for statical determinacy, it is necessary that:

$$3p = 3m + s \qquad (1.6)$$

otherwise, the number of redundants is given by:

$$r = 3m + s - 3p \qquad (1.7)$$

if s is not less than three and there are not more than three supporting elements at any joint. The method of inspection is, however, easily applied to

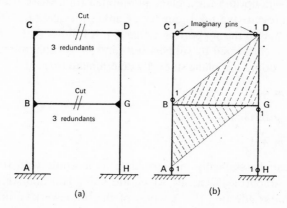

Fig. 1.4

frameworks of this kind. It consists of considering members cut and, or, considering joints pinned until the framework is statically determinate, when the inclusion of another pin-joint would result in its becoming a mechanism. This method is illustrated in Fig. 1.4 in which the plane framework shown has three components of support at each of its feet, A and H. Thus in Fig. 1.4(a) the redundants are removed by cutting members BG and CD while in Fig. 1.4(b) the redundants are removed by pinning at A, B, C, D, G and H in the manner shown. The cross-hatching in Fig. 1.4(b) is for the purpose of showing ABDG as a single member when there are pins at the six points.

Many engineers concerned with analysis of frameworks rarely apply formal tests of statical indeterminacy, partly since it is hardly necessary when the equilibrium approach to analysis is adopted (see Chapter 6) and partly because it is frequently possible to identify redundants by inspection.

1.3 Elasticity: Hooke's law and the principle of superposition

For practical purposes the common materials of engineering construction are elastic such that the deformation which they suffer under load disappears with removal of the load. Moreover, their elasticity is of a special kind in that the amount of any deformation is directly proportional to the load which

causes it. In other words, the strain of a structural member is directly proportional to its stress in accordance with the Hooke's law. Thus for a bar
or member in tension or compression the force T in it is related to its change
in length e such that $e = aT$ or $T = be$ where a ($=$ length/Young's modulus
$\times$ cross-sectional area) is the constant flexibility coefficient of the member
and b ($=1/a$) is the stiffness coefficient of the member.

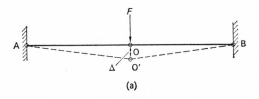

(a)

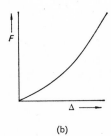

(b)

Fig. 1.5

This direct proportionality between strain and stress is called linear behaviour and with the justifiable assumption of small deformations, affords
considerable simplification in structural analysis, including application of the
principle of superposition. This principle is simply that when structural
behaviour is linear, the deflections or deformation due to a complex
system of loads is the sum of those effects caused by each component of the
loading acting separately upon the structure.

It is important to note that if the members of a structure are linearly
elastic the structure as a whole is not necessarily linearly elastic. Thus the
simple two-member pin-jointed system shown in Fig. 1.5(a) is non-linear for
vertical loading applied at O even when the members OA and OB are linear.
This is because significant change in geometry of the system is necessary to
provide a vertical component of force to resist the load. Hence the load/
deflection curve for vertical loading of O has the form shown in Fig. 1.5(b).
It is in circumstances such as this, where non-linearity arises or is modified
due to change in geometry (or large deflections), that analysis of quite simple
structures becomes complicated and some useful principles such as those of
superposition and complementary energy are invalid. This kind of behaviour

involving gross distortion cannot, however, be tolerated in engineering practice generally.

1.4 Strain energy, complementary energy and potential energy[1]

These quantities are often valuable for deriving relationships required in structural analysis.

Suppose that the elastic bar shown in Fig. 1.6 is subjected to a gradually increasing force F. Further, if it is assumed for the sake of generality, that the

Fig. 1.6

elasticity is non-linear, the force F might be related to the extension e of the bar in the manner described by the curve OA of Fig. 1.7. The area OAB below and to the right of the curve represents the work done by the force

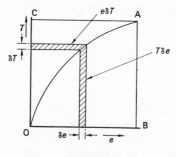

Fig. 1.7

upon the bar so that by the law of conservation of energy the strain energy stored in the bar is:

$$U = \int_0^e T \, de \tag{1.8}$$

where $T = F$ is the tension in the bar.
Clearly

$$\frac{dU}{de} = T = F \tag{1.9}$$

[1] See also bibliography, ref. 19.

Inspection of the area OAC above and to the left of the curve of Fig. 1.7 indicates, however, that it represents

$$C = \int_0^T e \, dT \tag{1.10}$$

a quantity which Engesser called the 'complementary energy'. Now:

$$\frac{dC}{dT} = e \tag{1.11}$$

It is important to note that the concept of complementary energy is artificial and has no physical status.

Strain energy is useful in theory of structures for deriving conditions of equilibrium while complementary energy enables conditions of compatibility of strain and deflection relationships to be derived. It is easy to verify that when the elasticity is linear the strain and complementary energies of a system are equal in magnitude though separately defined for the purpose of mathematical operations. Thus integrations of the kind of equation (1.8) for expressing the strain energy should always be performed by first expressing T as a function of e by means of the law of elasticity while those of equation (1.10) for expressing the complementary energy should always be performed by expressing e as a function of T.

The strain energy of a framework consisting of connected, axially loaded bars or members is given by adding the strain energies of the separate members, thus:

$$U = \sum_{}^{m} \int_0^{e_i} T_i \, de_i \tag{1.12}$$

where m is the number of members or bars; T_i and e_i are the axial force and change in length, respectively, of the ith member.

By the law of conservation of energy the strain energy is equal to the work done by the applied loads

$$U = W = \sum_{}^{n} \int_0^{\Delta_j} F_j \, d\Delta_j \tag{1.13}$$

where n is the number of components of load; F_j and Δ_j are the jth load component and deflection of the framework in the line of action of the jth load component respectively. Therefore:

$$U = \sum_{}^{m} \int_0^{e_i} T_i \, de_i = \sum_{}^{n} \int_0^{\Delta_j} F_j \, d\Delta_j \tag{1.14}$$

or using the law of linear elasticity and putting $T_i = b_i e_i$:

$$U = \tfrac{1}{2} \sum_{}^{m} b_i e_i^2 = \sum_{}^{n} \int_0^{\Delta_j} F_j^- d\Delta_j \tag{1.15}$$

9

Similarly, for complementary energy and work of the framework:

$$C = \sum_{}^{m} \int_{0}^{T_i} e_i \, dT_i = \sum_{}^{n} \int_{0}^{F_j} \Delta_j \, dF_j \tag{1.16}$$

or using the law of linear elasticity and putting $e_i = a_i T_i$ (where $a_i = 1/b_i$):

$$C = \tfrac{1}{2} \sum_{}^{m} a_i T_i^2 = \sum_{}^{n} \int_{0}^{F_j} \Delta_j \, dF_j \tag{1.17}$$

Equation (1.16) is immediately acceptable for linear elasticity owing to the equality of strain and complementary energy for this particular and common type of elasticity. The equation may be shown by the principle of virtual work to be valid in general provided that only small deformations or deflections of structures are considered. This restriction is trivial in relation to practical structures which without exception must carry loads without large deflection and consequent gross distortion geometrically.

For structural members in bending the strain and complementary energies are as follows[1]:

$$U = \int_{0}^{l} \int_{0}^{\phi} M \, d\phi \, dx \tag{1.18}$$

where ϕ is the (small) angle of bending per unit length of the member due to a bending moment M, x is the distance along the member from the origin at one end and l is the length of the member; and:

$$C = \int_{0}^{l} \int_{0}^{M} \phi \, dM \, dx \tag{1.19}$$

If the elasticity is linear, the simple (Bernoulli–Euler) theory of bending may be used when $\phi = M/EI$, $\delta\phi = \delta M/EI$ and $M = EI(d^2\Delta/dx^2)$ at any point distant x from the origin at which the flexural rigidity is EI and the deflection is Δ. Substituting accordingly in equations (1.18) and (1.19) gives for linearly elastic members:

$$U = \tfrac{1}{2} \int_{0}^{l} EI\left(\frac{d^2\Delta}{dx^2}\right)^2 dx \tag{1.20}$$

and

$$C = \tfrac{1}{2} \int_{0}^{l} \frac{M^2}{EI} \, dx \tag{1.21}$$

where $C = U$ numerically.

Therefore, for linear structures whose members resist loads in bending, equations (1.15) and (1.17) may be rewritten to include summations of quantities such as those of equations (1.20) and (1.21), respectively, for the strain and complementary energies of the members.

[1] See bibliography, refs. 2 and 19.

Potential energy,[1] the energy which a system possesses by virtue of its position and configuration is conveniently illustrated by considering a mass M under the influence of gravity suspended from a rigid anchorage by an elastic bar whose mass is negligible (Fig. 1.8). The potential energy of the motionless system including the bar and the mass consists of two components,

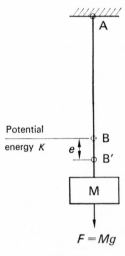

Fig. 1.8

the strain energy of the bar and the energy of the mass in the gravitational field. Thus the potential energy V of the system is:

$$V = U - Mge + k \tag{1.22}$$

where $U = \int_0^e T\,de$ is the strain energy of the bar, e is the total extension of the bar due to the force of gravity on the mass M, g is the acceleration due to gravity and k is the constant gravitational potential energy of M in its position of transfer to the bar.

Clearly:

$$\frac{dV}{de} = \frac{dU}{de} - Mg \tag{1.23}$$

but $dU/de = T$ and for equilibrium of the system $T = Mg$ so that, for equilibrium:

$$\frac{dV}{de} = 0 = \frac{dU}{de} - Mg \tag{1.24}$$

that is, the potential energy of a conservative system is stationary for a state of equilibrium.

[1] See bibliography, ref. 19.

11

It is important to note that the potential energy term relating to the mass of a system is not related to the work done on the elastic portion of the system. Thus it is assumed that the system is completed by using an external agency to introduce the mass (or load) and that this is done in such a way that no motion of the system ensues. In other words the external agency does work in introducing the mass (in a gravitational field) and the sum of this work and the work done on the elastic portion or structure of the system is equal to the loss of potential energy of the mass (*Mge*) as its weight is transferred to the structure.

The total potential of a system consisting of an elastic structure and its loading thus has the form:

$$V = U - \sum_{}^{n} F_j \Delta_j + k \tag{1.25}$$

assuming that the applied loads F_j ($j = 1, 2, \ldots, n$) are derived from the influence of gravity. The conditions for stationary potential energy are now:

$$\frac{\partial V}{\partial \Delta_j} = \frac{\partial U}{\partial \Delta_j} - F_j = 0 \quad (j = 1, 2, \ldots, n) \tag{1.26}$$

being conditions of equilibrium of the structure. If the n deflection components include all of the degrees of freedom of deflection of the structure, then equation (1.26) represents a complete set of equations of equilibrium of the system. (In general some of the Δs will represent angular deflection and some of the Fs, couples.)

As well as providing means of deriving equations of equilibrium in a manner similar to that afforded by the equation of work and strain energy, this principle also provides a means of investigating the stability of equilibrium, depending on whether the stationary property represents a maximum or a minimum.

1.5 The principle of virtual work

The principle of virtual work, which is useful in theory of structures for calculating deflections of structures is one of the older principles of mechanics being used as long ago as the thirteenth century.[1] It consists essentially of the application of the law of conservation of energy when a statical system in stable equilibrium suffers any small, arbitrary displacement. It is entirely general and so may be applied to frameworks with non-linear elasticity. But it is particularly useful as a device for calculating individual flexibility coefficients of linear frameworks. The principle of virtual work may be considered as a consequence of the principle of stationary total potential energy (para. 1.4).

[1] See bibliography, ref. 19.

Suppose that a system of forces $F_1, F_2, \ldots, F_N$ acts upon a particle P and that their resultant is F_R, as shown in Fig. 1.9. Now, if under the influence of these forces, P moves through a small distance Δ_R in the direction of F_R such

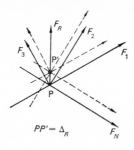

Fig. 1.9

that the magnitudes and directions of the forces are unchanged during the process, by the law of conservation of energy:

$$F_R\Delta_R = F_1\Delta_1 + F_2\Delta_2 + \cdots + F_N\Delta_N \tag{1.27}$$

where $\Delta_1, \Delta_2, \ldots, \Delta_N$ are the displacements in the lines of action of $F_1, F_2, \ldots, F_N$ respectively associated with Δ_R, that is the system of displacements $\Delta_R, \Delta_1, \Delta_2, \ldots, \Delta_N$ is geometrically compatible. Now if $F_R = 0$, the system of forces $F_1, F_2, \ldots, F_N$ is in equilibrium so that:

$$F_1\Delta_1 + F_2\Delta_2 + \cdots + F_N\Delta_N = 0 \tag{1.28}$$

The significance of this equation is that if a particle in equilibrium under the influence of a system of forces, is caused to suffer a small, arbitrary displacement (caused, say, by a temporary disturbance), the net work done by the forces is zero. From this it follows that if such a displacement is imagined to occur the expression for the work (virtual work) done by the forces can be equated to zero. This is the essence of the principle of virtual work which is concerned with imaginary but geometrically possible displacements of systems in equilibrium under the influence of forces and couples. It is important to note that, provided the geometry of the system (consisting of the particle and lines of action of the forces) is not violated, finite virtual displacements may be considered.

Application of the principle of virtual work to frameworks is conveniently described by reference to pin-jointed systems. At every joint of a framework of this kind the state of affairs illustrated in Fig. 1.9 exists: thus, the applied loads and the forces in the members at each joint correspond to a system of forces such as $F_1, F_2, \ldots, F_N$. Suppose the ith and jth joints of a pin-jointed framework are as shown in Fig. 1.10 connected by member q assumed to be in tension to the extent T_q and loaded by external forces F_{ix}, F_{iy}, F_{jx} and F_{jy},

13

respectively. If each of these joints is imagined to suffer a small, finite (virtual) displacement by Δ_{ix} and Δ_{iy} for joint i and Δ_{jx} and Δ_{jy} for joint j, which are compatible with displacements $(\Delta_q)_i$ at i and $(\Delta_q)_j$ at j in the direction and sense of T_q, the virtual work equations for the two joints are as follows:

$$F_{ix}\Delta_{ix} + F_{iy}\Delta_{iy} + T_q(\Delta_q)_i + \sum (T\Delta)_i = 0$$
$$F_{jx}\Delta_{jx} + F_{jy}\Delta_{jy} + T_q(\Delta_q)_j + \sum (T\Delta)_j = 0$$

$$(1.29)$$

where the summations embrace the forces in all members except q meeting

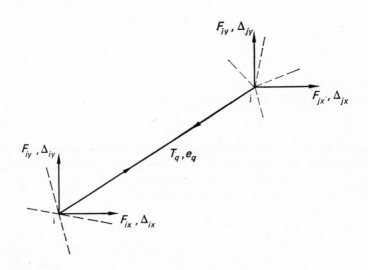

Fig. 1.10

at i and j respectively, which contribute to the maintenance of equilibrium at these joints. It follows that a pair of equations such as (1.29) can be written similarly for all pairs of joints and that if all such equations are added together, the following equation of virtual work is obtained:

$$\sum_{}^{p} F_{ix}\Delta_{ix} + \sum_{}^{p} F_{iy}\Delta_{iy} = \sum_{}^{m} T_q e_q$$

$$(1.30)$$

where p is the number of joints in the framework, m is the number of members or bars and e_q $(= -[(\Delta_q)_i + (\Delta_q)_j])$ is the virtual extension of the qth member. The reactions at supports may be included in this equation among the virtual work terms of the members of the framework. It must be emphasised that the Δs are a system of arbitrary, small, finite virtual displacements which are compatible with the geometry of the framework when it is loaded by forces F[1] which are in equilibrium with the forces in the members and supports T. Thus, the virtual displacements do not necessarily represent the system of displacements which are actually caused by the loads F. The choice of a system of virtual displacements is dictated entirely by expediency having regard to the objective in view.

If the members of the structure or framework resist loads primarily in bending then equation (1.30) becomes:

$$\sum_{}^{p} F_{ix}\Delta_{ix} + \sum_{}^{p} F_{iy}\Delta_{iy} = \sum_{}^{m} \int_{0}^{l} M\phi \, dx \qquad (1.31)$$

where ϕ is the virtual angular displacement per unit length at any point of a member whose length is l and M is the bending moment at that point, owing to the system of loads or external forces.

1.6 Continental, British and American influence in structural analysis

Study of the history of the development of structural analysis (Appendix I) indicates that the greatest influence has come from the Continent of Europe, beginning with the fundamental work of Navier (some of which is translated into English by Moseley) and continuing with that of Mohr, Engesser, Müller-Breslau and Ostenfeld. Mohr and Müller-Breslau made extensive use of the powerful principle of virtual work for calculating deflection of structures including deflection or flexibility coefficients and systematised a general approach to the analysis of linear statically indeterminate structures. This work set the style of continental structural analysis and except for refinement and devices employing the concepts of strain and complementary energy (the latter as an alternative to virtual work) there has been negligible change in principle. Thus, continental practice has been and is largely that of setting-up and solving formally the relevant simultaneous equations of a structure and is ideally suited to the programming of modern automatic computers.

Except for Maxwell's early work which preceded that of Mohr by a few years, and Southwell's relaxation (iterative) method, British engineers have exerted surprisingly little influence internationally on the general development of the subject until recent times. In fact, widespread sustained effort in analysis of statically indeterminate structures did not seemingly begin in the United Kingdom until about the end of the first world war. During the last twenty years British engineers have made significant contributions to modern theory

---

[1] If gross distortion under load is avoided in accordance with the requirements of engineering practice this is the same as the geometry of the unloaded framework, for practical purposes.

of structures including energy principles and were among the first to pro-gramme automatic computing machines for structural analysis. Moreover, they have made considerable contributions to the theory of ultimate load-carrying capacity of structures, an aspect which is outside the scope of this book.

American influence has been generally in respect of the use of scale models as an aid to structural analysis and the development of iterative techniques and approximate methods using energy principles. American engineers have, in recent years, played an important part in the use of automatic computing machines for structural analysis and in solving difficult problems posed by the structural design of aircraft. They are, moreover, acknowledged pioneers of the theory of stiffened suspension bridges.

It is emphasised that these remarks relate to the analysis of statically indeterminate structures.

EXERCISES

1 How many redundants are there in the plane pin-jointed framework shown in Fig. 1.11?

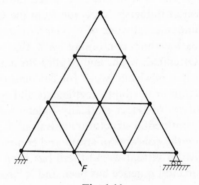

Fig. 1.11

Ans: 1

2 How many redundants are there in the plane rigidly jointed framework shown in Fig. 1.12?

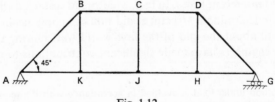

Fig. 1.12

Ans: 12

16

3 How many redundants are there in the freely jointed three-dimensional framework shown in Fig. 1.13?

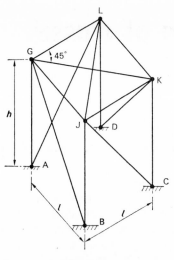

Fig. 1.13

Ans: 2

4 How many members must be removed from the framework shown in Fig. 1.13 for the system to become a mechanism? Explain why the answer to this question is apparently inconsistent with that of question 3 above.

Ans: 2

2

Deflection of trusses

2.1 Consideration of the analysis of deflection of elastic statically determinate trusses is necessary for the analysis of statically indeterminate structures of this kind in respect of both forces in members and the deflection of joints. Accordingly, this chapter is concerned with analytical (as distinct from graphical) methods of calculating the deflections of joints of trusses, including the basic geometrical method and those afforded by the concepts of complementary energy and virtual work. It is assumed throughout that deflections are sufficiently small that their effect on the geometry of a structure is negligible.

2.2 The basic geometrical method

This method which is laborious, is nevertheless instructive, and may be demonstrated by referring to a simple example. Suppose it is desired to calculate the two components Δ_v and Δ_h of the deflection of the 'free' joint O of the plane, pin-jointed elastic framework or truss (with two degrees of freedom) shown in Fig. 2.1 caused by the loading shown. The small changes in length of the two members OB and OA, e_1 and e_2, respectively, due to the loading are related to the deflection of the joint O as follows:

$$e_1 = -\Delta_h \cos \theta_1 - \Delta_v \sin \theta_1$$
$$e_2 = -\Delta_h \cos \theta_2 - \Delta_v \sin \theta_2 \tag{2.1}$$

simply by resolving Δ_h and Δ_v along the lines OA and OB of the members.[1] The angles θ_1 and θ_2 are constant for practical purposes since it is assumed that e_1, e_2, Δ_h and Δ_v are small. The minus signs in equation (2.1) indicate shortening of the members OA and OB for positive values of Δ_h and Δ_v and of the sines and cosines of the angles θ_1 and θ_2.

In order to make use of equations (2.1) it is clearly necessary that e_1 and e_2

---

[1] It is important to note that Δ_h and Δ_v cannot be obtained by resolving e_1 or e_2, i.e. $\Delta_h \neq -e_1 \cos \theta_1$; $\Delta_v \neq -e_2 \sin \theta_2$, because this process implies neglect of the small changes in θ_1 and θ_2 or rotations of OA and OB which are compatible with the straining of the structure.

are evaluated.[1] If the elasticity of the frameworks is linear and T_1 and T_2 are the forces (assumed to be tensile) in OB and OA respectively owing to the loading, then:

$$e_1 = \frac{T_1 l_1}{A_1 E_1}; \quad e_2 = \frac{T_2 l_2}{A_2 E_2} \tag{2.2}$$

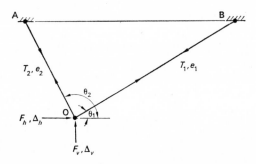

Fig. 2.1

where A_1 and A_2 are the cross-sectional areas and E_1 and E_2 are the Young's moduli of the members OB and OA respectively. Also, for equilibrium of O:

$$F_h + T_1 \cos \theta_1 + T_2 \cos \theta_2 = 0 \tag{2.3}$$
$$F_v + T_1 \sin \theta_1 + T_2 \sin \theta_2 = 0$$

Whence:

$$T_1 = -\left(\frac{F_h \sin \theta_2 - F_v \cos \theta_2}{\cos \theta_1 \sin \theta_2 - \sin \theta_1 \cos \theta_2}\right)$$

$$T_2 = -\left(\frac{F_h \sin \theta_1 - F_v \cos \theta_1}{\cos \theta_2 \sin \theta_1 - \sin \theta_2 \cos \theta_1}\right) \tag{2.4}$$

and

$$e_1 = -\left(\frac{F_h \sin \theta_2 - F_v \cos \theta_2}{\cos \theta_1 \sin \theta_2 - \sin \theta_1 \cos \theta_2}\right) \frac{l_1}{A_1 E_1}$$

$$e_2 = -\left(\frac{F_h \sin \theta_1 - F_v \cos \theta_1}{\cos \theta_2 \sin \theta_1 - \sin \theta_2 \cos \theta_1}\right) \frac{l_2}{A_2 E_2} \tag{2.5}$$

[1] The procedure here is based on e_1 and e_2 being due to load effects only: they may, however, be due in general to temperature variations also, in which event $e_1 = (T_1 l_1 / A_1 E_1) + \lambda_1$ and $e_2 = (T_2 l_2 / A_2 E_2) + \lambda_2$, where λ_1 and λ_2 represent the thermal effects. The resulting values of Δ_h and Δ_v are then due to both effects combined.

Returning now to equations (2.1), Δ_h and Δ_v may be expressed in terms of e_1 and e_2 as follows:

$$\Delta_h = +\left(\frac{e_1 \sin \theta_2 - e_2 \sin \theta_1}{\cos \theta_1 \sin \theta_2 - \sin \theta_1 \cos \theta_2}\right)$$

$$\Delta_v = +\left(\frac{e_1 \cos \theta_2 - e_2 \cos \theta_1}{\sin \theta_1 \cos \theta_2 - \cos \theta_1 \sin \theta_2}\right)$$

(2.6)

and by substituting for e_1 and e_2 from equations (2.5), Δ_h and Δ_v may be evaluated.

A more complicated example is provided by the plane, pin-jointed truss on rigid supports shown in Fig. 2.2. It has nine degrees of freedom represented

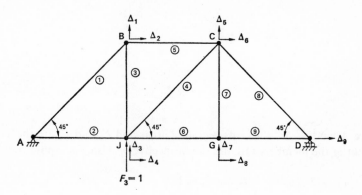

Fig. 2.2

by $\Delta_1, \Delta_2, \ldots, \Delta_9$ and nine members numbered as illustrated. The conditions of compatibility of strain and deflection, that is, the relationships between the changes in length of the members $e_1, e_2, \ldots, e_9$ and the components of deflection of the joints, $\Delta_1, \Delta_2, \ldots, \Delta_9$ are obtained by resolving the latter in the lines of the members, as follows:

$$e_1 = (\Delta_1 + \Delta_2)/\sqrt{2}$$
$$e_2 = \Delta_4$$
$$e_3 = (\Delta_1 - \Delta_3)$$
$$e_4 = [(\Delta_5 + \Delta_6) - (\Delta_3 + \Delta_4)]/\sqrt{2}$$
$$e_5 = (\Delta_6 - \Delta_2)$$
$$e_6 = (\Delta_8 - \Delta_4)$$
$$e_7 = (\Delta_5 - \Delta_7)$$
$$e_8 = (\Delta_5 + \Delta_9 - \Delta_6)/\sqrt{2}$$
$$e_9 = (\Delta_9 - \Delta_8)$$

(2.7)

If, then, each horizontal and vertical member has a flexibility of a (i.e. $l/AE = a$) while each inclined member has a flexibility of $a\sqrt{2}$ and there is a load $F_3 = 1$ applied at joint J, the forces in the members of the framework and their changes in length are as shown in Table 2.1 (the former being found by the conditions of equilibrium of the joints).

TABLE 2.1

Member	Member force T^*	Change in length $e = $ flexibility $\times T$
1	$2\sqrt{2}/3$	$4a/3$
2	$-2/3$	$-2a/3$
3	$-2/3$	$-2a/3$
4	$-\sqrt{2}/3$	$-2a/3$
5	$2/3$	$2a/3$
6	$-1/3$	$-a/3$
7	0	0
8	$\sqrt{2}/3$	$2a/3$
9	$-1/3$	$-a/3$

* Tension positive

Substitution of the values of e from Table 2.1 in equations (2.7) followed by their solution simultaneously gives the following values for the deflection components of the joints:

$$\Delta_1 = 4(3\sqrt{2} + 2)a/9; \quad \Delta_2 = -8a/9; \quad \Delta_3 = 2(6\sqrt{2} + 7)a/9$$

$$\Delta_4 = -2a/3; \quad \Delta_5 = 2(3\sqrt{2} + 5)a/9; \quad \Delta_6 = -2a/9 \quad (2.8)$$

$$\Delta_7 = 2(3\sqrt{2} + 5)a/9; \quad \Delta_8 = -a; \quad \Delta_9 = -4a/3$$

This process, which may also be used for calculating the deflections of the joints of pin-jointed space frameworks, is clearly laborious and is only justifiable if the deflection of every joint of a framework is required. If, as is customary, one or two deflections are needed methods afforded by the concepts of complementary energy and virtual work, respectively, are much more convenient.

2.3 The method of complementary energy

The complementary energy (C) of a structure is defined in para. 1.4 where it is shown that its partial derivative with respect to any individual load gives the deflection of the structure in the line of action of that load. If, therefore, it is required to find the vertical component of the deflection of joint J (Δ_3) of the truss shown in Fig. 2.2 due to a load F_3 acting at joint J, then:

$$\Delta_3 = \frac{\partial C}{\partial F_3} \quad (2.9)$$

where

$$C = \sum_{0}^{m} \int_{0}^{T_i} e_i \, dT_i = \frac{a\sqrt{2}}{2} (T_1^2 + T_4^2 + T_8^2)$$
$$+ \frac{a}{2} (T_2^2 + T_3^2 + T_5^2 + T_6^2 + T_7^2 + T_9^2)$$

or   (2.10)

$$\delta C = \sum^{m} e_i \, \delta T_i = a\sqrt{2} \, (T_1 \, \delta T_1 + T_4 \, \delta T_4 + T_8 \, \delta T_8)$$
$$+ a(T_2 \, \delta T_2 + T_3 \, \delta T_3 + T_5 \, \delta T_5$$
$$+ T_6 \, \delta T_6 + T_7 \, \delta T_7 + T_9 \, \delta T_9)$$

for this linearly elastic structure, so that:

$$\Delta_3 = \frac{\partial C}{\partial F_3} = a\sqrt{2} \left(T_1 \frac{\partial T_1}{\partial F_3} + T_4 \frac{\partial T_4}{\partial F_3} + T_8 \frac{\partial T_8}{\partial F_3} \right)$$
$$+ a\left(T_2 \frac{\partial T_2}{\partial F_3} + T_3 \frac{\partial T_3}{\partial F_3} + T_5 \frac{\partial T_5}{\partial F_3} \right.$$
$$\left. + T_6 \frac{\partial T_6}{\partial F_3} + T_7 \frac{\partial T_7}{\partial F_3} + T_9 \frac{\partial T_9}{\partial F_3} \right) \quad (2.11)$$

This calculation is best performed by tabulation, as shown in Table 2.2.

TABLE 2.2

Member	Flexibility	Member force T	$\dfrac{dT}{dF_3}$	$T\dfrac{dT}{dF_3}$	Member flexibility $\times T\dfrac{dT}{dF_3}$
1	$a\sqrt{2}$	$2\sqrt{2}\,F_3/3$	$2\sqrt{2}/3$	$8F_3/9$	$8\sqrt{2}\,aF_3/9$
2	a	$-2F_3/3$	$-2/3$	$4F_3/9$	$4aF_3/9$
3	a	$-2F_3/3$	$-2/3$	$4F_3/9$	$4aF_3/9$
4	$a\sqrt{2}$	$-\sqrt{2}\,F_3/3$	$-\sqrt{2}/3$	$2F_3/9$	$2\sqrt{2}\,aF_3/9$
5	a	$2F_3/3$	$2/3$	$4F_3/9$	$4aF_3/9$
6	a	$-F_3/3$	$-1/3$	$F_3/9$	$aF_3/9$
7	a	0	0	0	0
8	$a\sqrt{2}$	$\sqrt{2}\,F_3/3$	$\sqrt{2}/3$	$2F_3/9$	$2\sqrt{2}\,aF_3/9$
9	a	$-F_3/3$	$-1/3$	$F_3/9$	$aF_3/9$

$$\sum = 2aF_3(7 + 6\sqrt{2})/9$$

Thus:

$$\Delta_3 = 2aF_3(7 + 6\sqrt{2})/9 \quad\quad\quad (2.12)$$

which is in agreement with the value given in equations (2.8) when $F_3 = 1$.

If, however, the vertical deflection Δ_7 of joint G is required due to the load F_3 it is necessary to consider at first the action of a load F_7 at G in addition to F_3 at J. Then:

$$\left(\frac{\partial C}{\partial F_7} \right)_{F_7 = 0} = \Delta_7 \quad\quad\quad (2.13)$$

That is, the process is carried out to include F_7 and finally F_7 is given zero value. The calculation is shown in Table 2.3.

TABLE 2.3

Member	Member force T	$\dfrac{\partial T}{\partial F_7}$	$\left(T\dfrac{\partial T}{\partial F_7}\right)_{F_7=0}$	Member flexibility $\times\left(T\dfrac{\partial T}{\partial F_7}\right)_{F_7=0}$
1	$(2\sqrt2\,F_3+\sqrt2\,F_7)/3$	$\sqrt2/3$	$4F_3/9$	$4\sqrt2\,aF_3/9$
2	$-(2F_3+F_7)/3$	$-1/3$	$2F_3/9$	$2aF_3/9$
3	$-(2F_3+F_7)/3$	$-1/3$	$2F_3/9$	$2aF_3/9$
4	$\sqrt2(-F_3+F_7)/3$	$\sqrt2/3$	$-2F_3/9$	$-2\sqrt2\,aF_3/9$
5	$(2F_3+F_7)/3$	$1/3$	$2F_3/9$	$2aF_3/9$
6	$-(F_3+2F_7)/3$	$-2/3$	$2F_3/9$	$2aF_3/9$
7	$-F_7$	-1	0	0
8	$\sqrt2(F_3+2F_7)/3$	$2\sqrt2/3$	$4F_3/9$	$4\sqrt2\,aF_3/9$
9	$-(F_3+2F_7)/3$	$-2/3$	$2F_3/9$	$2aF_3/9$

$$\Sigma = 2aF_3(5+3\sqrt2)/9$$

Thus:

$$\Delta_7 = 2aF_3(5 + 3\sqrt2)/9 \tag{2.14}$$

in accordance with the value given in equations (2.8) when $F_3 = 1$.

This is a general procedure which may be used to calculate any component of deflection of a joint of a truss. It follows from the general form:

$$\frac{\partial C}{\partial F_j} = \Delta_j \quad (j = 1, 2, \ldots, n) \tag{2.15}$$

for a structure with n degrees of freedom. If some of the n forces F are zero this is taken into account after the derivatives of equation (2.15) have been obtained.

2.4 The method of virtual work

The principle of virtual work may be used as an alternative to the complementary energy method for calculating individual deflection components of the joints of a framework due to specified loading. If the forces T'_i ($i = 1, 2, \ldots, m$) in the m members and supporting elements of a framework are in equilibrium with any system of external forces or loads F'_j ($j = 1, 2, \ldots, n$) then for *any* system of small displacements e_i ($i = 1, 2, \ldots, m$), Δ_j ($j = 1, 2, \ldots, n$) which are compatible with the geometry of the loaded system and which are *imagined* to be introduced:

$$\sum_{}^{m} T'_i e_i = \sum_{}^{n} F'_j \Delta_j \tag{2.16}$$

in accordance with the discussion of para. 1.5.

23

Application of this principle may be demonstrated by referring again to the rigidly supported framework of Fig. 2.2. Since the components of deflection of joints depend upon the changes in length of the members due to specified loading, it is necessary to calculate the latter first of all. Then to obtain any one of the nine components of deflection of the joints due to the specified loading (virtual) displacements equal to the actual changes in length of the members are imagined to be given to the framework in equilibrium with an appropriately chosen external force of arbitrary magnitude. Thus, in order to find, say the vertical deflection Δ_7 of the joint G due to a load F_3 at J, the appropriate external force is an arbitrary vertical force F_7' at G in equilibrium with forces in the members T_i' ($i = 1, 2, \ldots, 9$). If, therefore, virtual displacements equal to the actual displacements e_i ($i = 1, 2, \ldots, 9$); Δ_j ($j = 1, 2, \ldots, 9$) are given to the system of forces in equilibrium F_7'; T_i' ($i = 1, 2, \ldots, 9$) the equation of virtual work is:

Compatible virtual displacements

$$F_7'\Delta_7 = \sum_{}^{9} T_i'e_i \tag{2.17}$$

Forces in equilibrium

whence

$$\Delta_7 = \frac{1}{F_7'} \sum^{9} T_i'e_i \tag{2.18}$$

being the vertical component of deflection of joint G due to a load F_3 at J. This calculation is shown in detail in Table 2.4 and for this purpose it is convenient to choose a value of $F_7' = 1$.

TABLE 2.4

Member	Force in member T due to F_3	Change in length e = Flexibility × T	Force in member T' due to $F_7' = 1$	$T'e$
1	$2\sqrt{2}\,F_3/3$	$4aF_3/3$	$\sqrt{2}/3$	$4\sqrt{2}\,aF_3/9$
2	$-2F_3/3$	$-2aF_3/3$	$-1/3$	$2aF_3/9$
3	$-2F_3/3$	$-2aF_3/3$	$-1/3$	$2aF_3/9$
4	$-\sqrt{2}\,F_3/3$	$-2aF_3/3$	$\sqrt{2}/3$	$-2\sqrt{2}\,aF_3/9$
5	$2F_3/3$	$2aF_3/3$	$1/3$	$2aF_3/9$
6	$-F_3/3$	$-aF_3/3$	$-2/3$	$2aF_3/9$
7	0	0	-1	0
8	$\sqrt{2}F_3/3$	$2aF_3/3$	$2\sqrt{2}/3$	$4\sqrt{2}\,aF_3/9$
9	$-F_3/3$	$-aF_3/3$	$-2/3$	$2aF_3/9$

$$\Sigma = 2aF_3(5+3\sqrt{2})/9$$

Therefore:

$$\Delta_7 = 2aF_3(5+3\sqrt{2})/9 \tag{2.19}$$

in accordance with equations (2.8) and (2.14).

The same procedure is necessary regardless of the loading of the structure for which an individual component of deflection is required. Thus if the loading consisted of $F_1, F_2, \ldots, F_9$ the first step in finding any component of deflection of a joint is the calculation of the (small) changes in length e of the members due to the specified loading, being compatible with the deflections of the joints. These values of e are then used as virtual displacements of the system of member forces T' in equilibrium with an arbitrary external force acting at the joint whose component of deflection is required and in the line of that component of deflection.

The calculation of the deflection of a structure loaded by a single force in the direction of the line of action of that force is a particular situation in which the requisite force and virtual displacement systems correspond to the real systems of the loaded state. The relevant virtual work equation is now:

$$F_j \Delta_j = \sum_{}^{m} T_i e_i \qquad (2.20)$$

where F_j is the single load, and this result may be obtained alternatively by direct application of the law of conservation of energy.[1] For a linearly elastic structure this gives:

$$\tfrac{1}{2} F_j \Delta_j = \tfrac{1}{2} \sum_{}^{m} T_i e_i \qquad (2.21)$$

whence:

$$F_j \Delta_j = \sum_{}^{m} T_i e_i \qquad (2.22)$$

The method of virtual work, in common with that of complementary energy, may be used to calculate deflections of non-linearly elastic structures also, provided the assumption that deflections are small is noted. The procedure is precisely the same as that described herein, except that the appropriate law of elasticity must be used to obtain the changes in length e of the structural members.

2.5 Example of the calculation of deflections by complementary energy and virtual work

The linearly elastic, plane, pin-jointed framework shown in Fig. 2.3 is a Warren truss, all of the members of which are identical with a flexibility of a. It has seven degrees of freedom and seven members. It is required to find the horizontal deflection of D, Δ_5, due to a load F_6, and a non-uniform thermal effect such that members 1, 2 and 3 each suffer a small increase in length from this source of λ while the remainder suffer an increase in length of 2λ.

---

[1] See also para. 4.6.

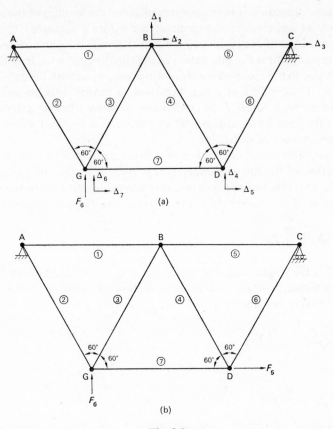

Fig. 2.3

For the purpose of using the complementary energy method it is necessary to introduce a load F_5 in addition to F_6 and to obtain the required deflections due to F_6 and the thermal effect by putting $F_5 = 0$ finally.   Thus:

$$\delta C = \sum_{}^{7} (\lambda_i + a_i T_i)\, \delta T_i \tag{2.23}$$

and

$$\Delta_5 = \left(\frac{\partial C}{\partial F_5}\right)_{F_5 = 0} = \left[\sum_{}^{7} (\lambda_i + a_i T_i)\frac{\partial T_i}{\partial F_5}\right]_{F_5 = 0} \tag{2.24}$$

The operations of equation (2.24) is conveniently performed in tabular form, as shown in Table 2.5.

Thus, owing to a load F_6 and the thermal effect:

$$\Delta_5 = 9\lambda/4 + aF_6\sqrt{3}/8 \tag{2.25}$$

The horizontal deflection Δ_7 of G may be calculated immediately, if desired, because it is simply Δ_5 plus the shortening of member DG(7) owing to the load F_6, i.e.

$$\Delta_7 = \Delta_5 + aF_6/2\sqrt{3} - 2\lambda = \lambda/4 + 7aF_6/8\sqrt{3} \qquad (2.26)$$

TABLE 2.5

Member	Member Force T	$\left(\dfrac{\partial T_i}{\partial F_5}\right)$	$\left[(\lambda_i + a_i T_i)\dfrac{\partial T_i}{\partial F_5}\right]_{F_5=0}$
1	$F_6\sqrt{3}/4 + 3F_5/4$	3/4	$3\lambda/4 + 3aF_6\sqrt{3}/16$
2	$-F_6\sqrt{3}/2 + F_5/2$	1/2	$\lambda/2 - 4aF_6\sqrt{3}/16$
3	$-F_6/2\sqrt{3} - F_5/2$	$-1/2$	$-\lambda/2 + 4aF_6/16\sqrt{3}$
4	$F_6/2\sqrt{3} + F_5/2$	1/2	$\lambda + 4aF_6/16\sqrt{3}$
5	$F_6/4\sqrt{3} + F_5/4$	1/4	$\lambda/2 + aF_6/16\sqrt{3}$
6	$-F_6/2\sqrt{3} - F_5/2$	$-1/2$	$-\lambda + 4aF_6/16\sqrt{3}$
7	$-F_6/2\sqrt{3} + F_5/2$	1/2	$\lambda - 4aF_6/16\sqrt{3}$
			$\Sigma = 9\lambda/4 + aF_6\sqrt{3}/8$

In order, now, to use the principle of virtual work to find Δ_5 it is necessary to use the displacements or changes in length of the members due to the actual loading, F_6, and the thermal effect as compatible virtual displacements for the system in equilibrium when there is an arbitrary force F_5 applied at D in the line of Δ_5. It is convenient for this purpose to put $F_5 = 1$. Then if the changes in length of the members due to the actual loading and thermal effect are e_1 ($i = 1, 2, \ldots, 7$) while the forces in the members due to a force $F_5 = 1$ at D are T'_i ($i = 1, 2, \ldots, 7$), by the principle of virtual work:

$$1\Delta_5 = \sum_{}^{7} T'_i e_i \qquad (2.27)$$

This calculation is conveniently carried out in tabular form as shown in Table 2.6.

TABLE 2.6

Member	Member Force, T due to F_6	Change in length of member, e due to F_6 and thermal effect	Member Force, T' due to $F_5 = 1$	$T'e$
1	$F_6\sqrt{3}/4$	$\lambda + aF_6\sqrt{3}/4$	3/4	$3\lambda/4 + 3aF_6\sqrt{3}/16$
2	$-F_6\sqrt{3}/2$	$\lambda - aF_6\sqrt{3}/2$	1/2	$\lambda/2 - 4aF_6\sqrt{3}/16$
3	$-F_6/2\sqrt{3}$	$\lambda - aF_6/2\sqrt{3}$	$-1/2$	$-\lambda/2 + 4aF_6/16\sqrt{3}$
4	$F_6/2\sqrt{3}$	$2\lambda + aF_6/2\sqrt{3}$	1/2	$\lambda + 4aF_6/16\sqrt{3}$
5	$F_6/4\sqrt{3}$	$2\lambda + aF_6/4\sqrt{3}$	1/4	$\lambda/2 + aF_6/16\sqrt{3}$
6	$-F_6/2\sqrt{3}$	$2\lambda - aF_6/2\sqrt{3}$	$-1/2$	$-\lambda + 4aF_6/16\sqrt{3}$
7	$-F_6/2\sqrt{3}$	$2\lambda - aF_6/2\sqrt{3}$	1/2	$\lambda - 4aF_6/16\sqrt{3}$
				$\Sigma = 9\lambda/4 + aF_6\sqrt{3}/8$

From Table 2.6 then:

$$\Delta_5 = 9\lambda/4 + aF_6\sqrt{3}/8 \tag{2.28}$$

in accordance with equation (2.25) obtained by the complementary energy method. It will be noted that the values in the fourth column of Table 2.5 are identical to those of the fifth column of Table 2.6.

2.6 Flexibility and stiffness coefficients of linearly elastic trusses

For any linearly elastic structure the deflection in any given direction at a point, caused by unit load applied anywhere in any direction, is called a flexibility coefficient. Using the framework shown in Fig. 2.2 as an example, the deflection components of joint B due to unit vertical load at that point are the flexibility coefficients a_{11} and a_{21}, respectively, where the first subscript describes the deflection component and the second subscript indicates the location of the unit load. Due to this same unit load the components of deflection of joint J are a_{31} and a_{41}, respectively, for joint C they are a_{51} and a_{61}, respectively, for joint G they are a_{71} and a_{81}, respectively, and for joint D the single component of deflection is a_{91}. Similarly, for unit horizontal load at B, the components of deflection of the joints of the structure are a_{12}, a_{22}, a_{32}, a_{42}, a_{52}, a_{62}, a_{72}, a_{82} and a_{92} respectively, and in general due to $F_j = 1$, $\Delta_i = a_{ij}$. Calculation of flexibility coefficients is achieved by using the complementary or virtual work methods (paras. 2.3 and 2.4). For example, by equations (2.8), (2.14) and (2.19) $\Delta_7 = 2a(5 + 3\sqrt{2})/9$, when $F_3 = 1$, i.e. $a_{73} = 2a(5 + 3\sqrt{2})/9$ ($= a_{37}$, see below).

On the basis of the principle of superposition for linearly elastic structures, the components of deflection of the joints of the structure due to general loading $F_1, F_2, \ldots, F_9$ may be expressed in terms of the loads by means of the relevant flexibility coefficients, as follows:

$$
\begin{aligned}
\Delta_1 &= a_{11}F_1 + a_{12}F_2 + a_{13}F_3 + \cdots + a_{19}F_9 \\
\Delta_2 &= a_{21}F_1 + a_{22}F_2 + a_{23}F_3 + \cdots + a_{29}F_9 \\
\Delta_3 &= a_{31}F_1 + a_{32}F_2 + a_{33}F_3 + \cdots + a_{39}F_9 \\
&\vdots \\
\Delta_9 &= a_{91}F_1 + a_{92}F_2 + a_{93}F_3 + \cdots + a_{99}F_9
\end{aligned}
\tag{2.29}
$$

Transformation of these equations gives the loads to cause specified deflections of the joints of the structure in terms of those deflections and the relevant stiffness coefficients as follows:

$$
\begin{aligned}
F_1 &= b_{11}\Delta_1 + b_{12}\Delta_2 + b_{13}\Delta_3 + \cdots + b_{19}\Delta_9 \\
F_2 &= b_{21}\Delta_1 + b_{22}\Delta_2 + b_{23}\Delta_3 + \cdots + b_{29}\Delta_9 \\
F_3 &= b_{31}\Delta_1 + b_{32}\Delta_2 + b_{33}\Delta_3 + \cdots + b_{39}\Delta_9 \\
&\vdots \\
F_9 &= b_{91}\Delta_1 + b_{92}\Delta_2 + b_{93}\Delta_3 + \cdots + b_{99}\Delta_9
\end{aligned}
\tag{2.30}
$$

Accordingly, any stiffness coefficient, say b_{ij}, may be defined as the external force necessary to cause Δ_i to be zero when $\Delta_j = 1$ owing to a force b_{jj} and every other deflection component is zero due to forces b_{1j}, b_{2j}, etc. It is im-

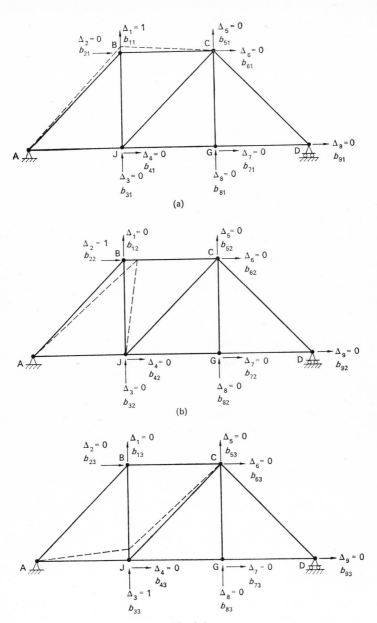

Fig. 2.4

portant to note that these concepts are applicable for both statically determinate and indeterminate linear structures.

The stiffness coefficients of a framework are easily calculated directly without recourse to any special device. Thus to find $b_{11}, b_{21}, b_{31}, \ldots, b_{91}$ for the framework shown in Fig. 2.2 it is simply necessary to calculate the forces which must be applied to the joints to cause $\Delta_1 = 1$ and $\Delta_2 = \Delta_3 = \cdots = \Delta_9 = 0$, as shown in Fig. 2.4(a). For this system of deflections:

$$e_1 = 1/\sqrt{2}; \quad e_2 = 0; \quad e_3 = 1; \quad e_4 = 0;$$
$$e_5 = 0; \qquad e_6 = 0; \quad e_7 = 0; \quad e_8 = 0; \quad e_9 = 0 \tag{2.31}$$

whence:

$$T_1 = 1/2a; \quad T_2 = 0; \quad T_3 = 1/a; \quad T_4 = 0;$$
$$T_5 = 0; \qquad T_6 = 0; \quad T_7 = 0; \qquad T_8 = 0; \quad T_9 = 0 \tag{2.32}$$

and by resolution at the joints:

$$b_{11} = T_1/\sqrt{2} + T_3 = 1/2\sqrt{2}\,a + 1/a = (1 + 2\sqrt{2})/2\sqrt{2}\,a$$
$$b_{21} = T_1/\sqrt{2} = 1/2\sqrt{2}\,a$$
$$b_{31} = -T_3 = -1/a \tag{2.33}$$
$$b_{41} = 0 = b_{51} = b_{61} = b_{71} = b_{81} = b_{91}$$

Similarly, to find $b_{12}, b_{22}, b_{32}, \ldots, b_{92}$, the forces which must be applied to the joints to cause $\Delta_2 = 1; \Delta_1 = \Delta_3 = \cdots = \Delta_9 = 0$, as shown in Fig. 2.4(b), are calculated as follows:

$$e_1 = 1/\sqrt{2}; \quad e_2 = 0; \quad e_3 = 0; \quad e_4 = 0; \quad e_5 = -1;$$
$$e_6 = 0; \qquad e_7 = 0; \quad e_8 = 0; \quad e_9 = 0 \tag{2.34}$$

whence:

$$T_1 = 1/2a; \quad T_2 = 0; \quad T_3 = 0; \quad T_4 = 0; \quad T_5 = -1/a$$
$$T_6 = 0; \qquad T_7 = 0; \quad T_8 = 0; \quad T_9 = 0 \tag{2.35}$$

and by resolution at the joints:

$$b_{12} = T_1/\sqrt{2} = 1/2\sqrt{2}\,a$$
$$b_{22} = T_1/\sqrt{2} - T_5 = 1/2\sqrt{2}\,a + 1/a = (1 + 2\sqrt{2})/2\sqrt{2}\,a$$
$$b_{32} = 0 = b_{42} = b_{52} = b_{72} = b_{82} = b_{92} \tag{2.36}$$
$$b_{62} = T_5 = -1/a$$

Again, for $b_{13}, b_{23}, b_{33}, \ldots, b_{93}$, $\Delta_3 = 1$ and $\Delta_1 = \Delta_2 = \Delta_4 = \cdots = \Delta_9 = 0$, as shown in Fig. 2.4(c) and:

$$e_1 = 0; \quad e_2 = 0; \quad e_3 = -1; \quad e_4 = -1/\sqrt{2}; \quad e_5 = 0;$$
$$e_6 = 0; \quad e_7 = 0; \quad e_8 = 0; \qquad e_9 = 0 \tag{2.37}$$

whence:

$$T_1 = 0; \quad T_2 = 0; \quad T_3 = -1/a; \quad T_4 = -1/2a;$$
$$T_5 = 0; \quad T_6 = 0; \quad T_7 = 0; \qquad T_8 = 0; \qquad T_9 = 0 \tag{2.38}$$

Therefore:

$$b_{13} = -1/a; \quad b_{23} = 0; \quad b_{33} = 1/a + 1/2\sqrt{2}\,a = (1 + 2\sqrt{2})/2\sqrt{2}\,a$$
$$b_{43} = 1/2\sqrt{2}\,a; \quad b_{53} = -1/2\sqrt{2}\,a; \quad b_{63} = -1/2\sqrt{2}\,a; \tag{2.39}$$
$$b_{73} = 0; \quad b_{83} = 0; \quad b_{93} = 0$$

By repeating the same procedure for $\Delta_4 = 1$; $\Delta_1 = \Delta_2 = \Delta_3 = \Delta_5 = \cdots$ $= \Delta_9 = 0$: $\Delta_5 = 1$; $\Delta_1 = \Delta_2 = \Delta_3 = \Delta_4 = \Delta_6 = \cdots = \Delta_9 = 0$: $\Delta_6 = 1$; $\Delta_1 = \Delta_2 = \cdots = \Delta_5 = \Delta_7 = \Delta_8 = \Delta_9 = 0$: $\Delta_7 = 1$; $\Delta_1 = \Delta_2 = \cdots = \Delta_6$ $= \Delta_8 = \Delta_9 = 0$: $\Delta_8 = 1$; $\Delta_1 = \Delta_2 = \cdots = \Delta_7 = \Delta_9 = 0$: $\Delta_9 = 1$; Δ_1 $= \Delta_2 = \cdots = \Delta_8 = 0$, respectively in succession the remainder of the eighty-one stiffness coefficients for the structure may be calculated. It will be observed finally that any 'cross coefficient' such as b_{ij} is identical to b_{ji}. This is a special feature of the behaviour of linear systems and is known as their reciprocal property. The same property is manifest in the flexibility coefficients so that in general $a_{ij} = a_{ji}$. Recognition of the reciprocal property indicates that for a system with n degrees of freedom it is never necessary to calculate all n^2 of the stiffness or the corresponding flexibility coefficients. Instead calculation of only the n 'self' coefficients such as a_{ii} or b_{ii} and $n(n-1)/2$ cross coefficients, is necessary.

It is important to note that the individual stiffness coefficients are not related in a simple manner to the individual flexibility coefficients, i.e. $b_{ij} \neq 1/a_{ij}$.

The easily calculated stiffness coefficients are, however, of little value in matters concerning statically determinate frameworks. The calculation of individual flexibility coefficients is a relatively laborious process whatever device is used for the purpose, e.g. the complementary energy or virtual work methods. The value of the flexibility coefficients lies in their enabling deflections caused by any system of loading, or combinations of loading systems, to be calculated readily.

Finally, although in this treatment flexibility and stiffness coefficients are calculated in respect of orthogonal components of deflection of joints, other sets of these coefficients exist for any specified lines or directions taken through the joints of a framework.

2.7 Example of the calculation of flexibility and stiffness coefficients of a truss

The plane cantilever truss shown in Fig. 2.5 has linearly elastic members such that the flexibility of the inclined members is a, that of the horizontal members is $4a/5$ and that of the vertical members is $3a/5$. (That is, as though the members are of the same material and cross-sectional area, since then the relative flexibilities of the members are directly proportional to their lengths.) The truss has eight degrees of freedom, defined as shown in Fig. 2.5. It is required to find the flexibility of the truss in the lines BG and CE for vertical loading at D, that is, to find $a_{(BG)5}$ and $a_{(CE)5}$, and the flexibility $a_{(BG)(CE)}$ in the line BG for simultaneously applied equal and opposite forces at C and E acting in the line CE (as though there is a member CE in tension or compression). It is also required to find the stiffness coefficients of the truss.

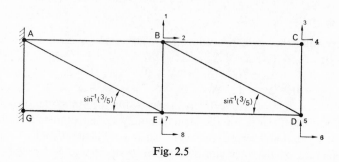

Fig. 2.5

In order to calculate the flexibility coefficients by means of the principle of virtual work it is necessary to determine the appropriate force and displacement systems. These are shown in Fig. 2.6 for this problem. Thus for $a_{(BG)5}$, the shortening of the line BG owing to $F_5 = 1$, application of the principle of virtual work to the force system in equilibrium of Fig. 2.6(b) using compatible virtual displacements equal to those caused by the loading of Fig. 2.6(a) gives:

$$1\Delta_{BG} = 1a_{(BG)5} = \sum_{}^{8} T'_i e_i \tag{2.40}$$

where the changes in length of the members e_i ($i = 1, 2, \ldots, 8$) are obtained by multiplying the forces in the members caused by $F_5 = 1$ by the respective flexibilities.

Again, for $a_{(CE)5}$, application of the principle of virtual work to the force system of Fig. 2.6(c) using the same compatible virtual displacements as for equation (2.40) gives:

$$1\Delta_{CE} = 1a_{(CE)5} = \sum_{}^{8} T''_i e_i \tag{2.41}$$

The processes of equations (2.40) and (2.41) are shown in Table 2.7. That is:

$$a_{(BG)5} = -72a/25 = a_{5(BG)} \tag{2.42}$$

$$a_{(CE)5} = -72a/25 = a_{5(CE)} \tag{2.43}$$

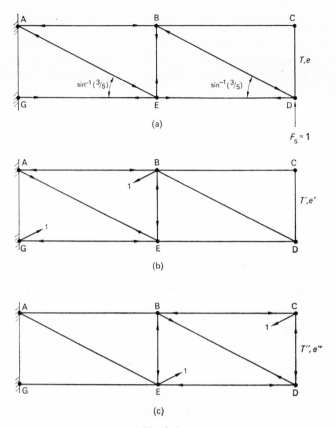

Fig. 2.6

TABLE 2.7

Member	T	e	T'	$T'e$	T''	$T''e$
AB	$-4/3$ *	$-16a/15$	$-4/5$	$64a/75$	0	0
AE	$-5/3$	$-5a/3$	1	$-5a/3$	0	0
BC	0	0	0	0	$-4/5$	0
BD	$-5/3$	$-5a/3$	0	0	1	$-5a/3$
BE	1	$3a/5$	$-3/5$	$-9a/25$	$-3/5$	$-9a/25$
CD	0	0	0	0	$-3/5$	0
DE	$4/3$	$16a/15$	0	0	$-4/5$	$-64a/75$
EG	$8/3$	$32a/15$	$-4/5$	$-128a/75$	0	0
				$\sum = -72a/25$		$\sum = -72a/25$

* Minus sign indicates compression.

2*

It is incidental that $a_{(BG)5} = a_{(CE)5}$ and is particular to this example. The minus signs indicate that the changes in the lengths of the lines BG and CE are in the opposite sense to that of the unit forces applied in those lines as shown in Fig. 2.6(b) and 2.6(c), respectively; that is, the coefficients represent an increase in the lengths of the lines due to $F_5 = 1$. Thus, for, say, $F_5 = -25$ units of force:

$$\Delta_{BG} = -25a_{(BG)5} = -25 \times -72a/25 = 72a \qquad (2.44)$$

$$\Delta_{CE} = -25a_{(CE)5} = -25 \times -72a/25 = 72a \qquad (2.45)$$

In order now to calculate $a_{(BG)(CE)}$, the virtual work of the force system in equilibrium of Fig. 2.6(b) for virtual displacements equal to the displacements caused by the force system of Fig. 2.6(c) gives:

$$1\Delta_{BG} = 1a_{(BG)(CE)} = \sum_{i}^{8} T'_i e''_i \qquad (2.46)$$

There is no need to deal with this equation by tabulation because all except one product $T'_i e''_i$ (that in respect of member BE) are zero: Thus:

$$a_{(BG)(CE)} = -3/5 \times -3/5 \times 3a/5 = 27a/125 \qquad (2.47)$$

It is easily verified that $a_{(BG)5} = a_{5(BG)}$, $a_{(CE)5} = a_{5(CE)}$ and $a_{(BG)(CE} = a_{(CE)(BG)}$. For example, by the principle of virtual work:

$$1a_{5(BG)} = \sum_{i}^{8} T_i e'_i \qquad (2.48)$$

but $e'_i = a_i T'_i$, a_i being the relevant member flexibility and if this substitution is made in equations (2.40) and (2.48) their identity is immediately apparent.

In order now to find the stiffness coefficients of the truss, unit displacements are considered in turn in each of the eight degrees of freedom and the external forces required to cause these eight separate conditions calculated. This process is shown in Tables 2.8 and 2.9.

For equilibrium of the joints of the framework having regard to the forces in the members given in Table 2.8, the external forces shown in Table 2.9 are necessary, being the required stiffness coefficients.

It is important to note that having calculated the eight coefficients b_{11}, b_{21}, etc. of the first row of Table 2.9 only seven coefficients $b_{22}, b_{32}, \ldots, b_{82}$ need be calculated for the second row, only six $b_{33}, \ldots, b_{83}$ need be calculated for the third row, five for the fourth row, four for the fifth row, three for the sixth row, two for the seventh row and one, b_{88} for the eighth row. This is because the reciprocal property of stiffness coefficients whence $b_{ij} = b_{ji}$ may be used to complete the table, i.e. to the left of the diagonal containing $b_{11}, b_{22}, \ldots, b_{88}$.

2.8 The reciprocal theorem

The reciprocal theorem is valid for systems whose behaviour may be specified by linear relationships. It was first expounded explicitly by Maxwell in 1864

TABLE 2.8

Change in length when:

Member	$\Delta_1 = 1$	$\Delta_2 = 1$	$\Delta_3 = 1$	$\Delta_4 = 1$	$\Delta_5 = 1$	$\Delta_6 = 1$	$\Delta_7 = 1$	$\Delta_8 = 1$
AB	0	1	0	0	0	0	0	0
BC	0	-1	0	1	0	0	0	0
BD	3/5	-4/5	0	0	-3/5	4/5	0	0
BE	1	0	0	0	0	0	-1	0
CD	0	0	-1	0	-1	0	0	0
DE	0	0	0	0	0	1	0	-1
AE	0	0	0	0	0	0	-3/5	4/5
EG	0	0	0	0	0	0	0	1

Force in member when:

Member	$\Delta_1 = 1$	$\Delta_2 = 1$	$\Delta_3 = 1$	$\Delta_4 = 1$	$\Delta_5 = 1$	$\Delta_6 = 1$	$\Delta_7 = 1$	$\Delta_8 = 1$
AB	0	5/4a	0	5/4a	0	0	0	0
BC	0	-5/4a	0	5/4a	-3/5a	0	0	0
BD	3/5a	-4/5a	0	0	0	4/5a	0	0
BE	5/3a	0	0	0	0	0	-5/3a	0
CD	0	0	5/3a	0	-5/3a	0	0	0
DE	0	0	0	0	0	5/4a	0	-5/4a
AE	0	0	0	0	0	0	-3/5a	4/5a
EG	0	0	0	0	0	0	0	5/4a

TABLE 2.9

	Direction 1	Direction 2	Direction 3	Direction 4	Direction 5	Direction 6	Direction 7	Direction 8
$\Delta_1 = 1$*	b_{11} $5/3a + 9/25a$	b_{21} $-12/25a$	b_{31} 0	b_{41} 0	b_{51} $-9/25a$	b_{61} $12/25a$	b_{71} $-5/3a$	b_{81} 0
$\Delta_2 = 1$	b_{12} $-12/25a$	b_{22} $5/2a + 16/25a$	b_{32} 0	b_{42} $-5/4a$	b_{52} $12/25a$	b_{62} $-16/25a$	b_{72} 0	b_{82} 0
$\Delta_3 = 1$	b_{13} 0	b_{23} 0	b_{33} $5/3a$	b_{43} 0	b_{53} $-5/3a$	b_{63} 0	b_{73} 0	b_{83} 0
$\Delta_4 = 1$	b_{14} 0	b_{24} $-5/4a$	b_{34} 0	b_{44} $5/4a$	b_{54} 0	b_{64} 0	b_{74} 0	b_{84} 0
$\Delta_5 = 1$	b_{15} $-9/25a$	b_{25} $12/25a$	b_{35} $-5/3a$	b_{45} 0	b_{55} $5/3a + 9/25a$	b_{65} $-12/25a$	b_{75} 0	b_{85} 0
$\Delta_6 = 1$	b_{16} $12/25a$	b_{26} $-16/25a$	b_{36} 0	b_{46} 0	b_{56} $-12/25a$	b_{66} $5/4a + 16/25a$	b_{76} 0	b_{86} $-5/4a$
$\Delta_7 = 1$	b_{17} $-5/3a$	b_{27} 0	b_{37} 0	b_{47} 0	b_{57} 0	b_{67} 0	b_{77} $5/3a + 9/25a$	b_{87} $-12/25a$
$\Delta_8 = 1$	b_{18} 0	b_{28} 0	b_{38} 0	b_{48} 0	b_{58} 0	b_{68} $-5/4a$	b_{78} $-12/25a$	b_{88} $5/2a + 16/25a$

* All other deflections being maintained zero.

with reference to linearly elastic frameworks, though Clebsch had noted the reciprocal property of stiffness coefficients some two years earlier.[1] Later, in 1872, the Italian, Betti, dealt with the theorem independently in general terms. The theorem is manifest in the reciprocal property of the flexibility and stiffness coefficients, respectively, of linearly elastic systems. Thus, it is noted in para. 2.6 above that $a_{ij} = a_{ji}$ and $b_{ij} = b_{ji}$, respectively.

Stated verbally with reference to structures the theorem is that the deflection of a structure in a specified line i due to unit force acting in a specified line j is the same as the deflection in line j due to unit force acting in the line i. Alternatively, if unit deflection in the ith degree of freedom of a structure is caused by external forces such that the deflections in all other degrees of freedom are zero, the restraining force required in the jth degree of freedom is the same as that required in the ith degree of freedom when there is unit deflection in the jth degree of freedom only.

Verification of the theorem may be achieved in simple terms by application of the principle of virtual work to, say, a linearly elastic plane pin-jointed truss with n members. If unit force is applied in the ith degree of freedom (i.e. $F_i = 1$) causing forces T'_m ($m = 1, 2, \ldots, n$) in the members and corresponding changes in their length of $e'_m = a_m T'_m$ ($m = 1, 2, \ldots, n$) the deflection produced in the jth degree of freedom is $\Delta'_j = a_{ji}$. Then to calculate a_{ji} by the principle of virtual work it is necessary to consider the system in equilibrium when unit force acts in the jth degree of freedom causing forces T''_m ($m = 1, 2, \ldots, n$) in the members. Application of the principle of virtual work to the latter with virtual displacement equal to e'_m ($m = 1, 2, \ldots, n$) gives:

$$1\Delta'_j = 1a_{ji} = \sum_{n} T''_m e'_m = \sum_{n} T''_m T'_m a_m \tag{2.49}$$

But if the force system T'_m ($m = 1, 2, \ldots, n$) caused by $F_i = 1$ is considered to suffer virtual displacements $e''_m = a_m T''_m$ ($m = 1, 2, \ldots, n$) equal to the actual displacements which would be caused by $F_j = 1$, then:

$$1\Delta''_i = 1a_{ij} = \sum_{n} T'_m e''_m = \sum_{n} T'_m T''_m a_m \tag{2.50}$$

whence it is apparent from the identity of equations (2.49) and (2.50) that $a_{ij} = a_{ji}$.

Besides the saving in labour for the calculation of sets of flexibility and stiffness coefficients of a system, the reciprocal theorem affords useful theorems relating to influence lines for force and deflection (see para. 7.2).

2.9 Use of symmetry and anti-symmetry[2]

Consideration of symmetry and anti-symmetry is important to economy of effort when analysing linearly elastic structures for forces and deflections due to prescribed conditions of loading. For this purpose it is essential that the

---

[1] See Appendix I.
[2] This device is believed to be due to Bresse.

structure shall have an axis of symmetry in respect of both overall geometry and sections of members.

A symmetrical Warren truss is shown in Fig. 2.7(a) loaded symmetrically.

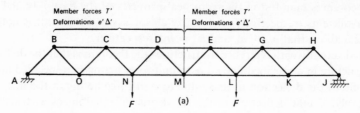

Member forces T'
Deformations $e' \, \Delta'$

Member forces T'
Deformations $e' \, \Delta'$

(a)

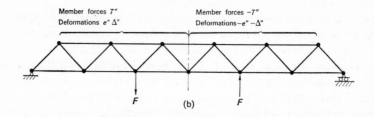

Member forces T''
Deformations $e'' \, \Delta''$

Member forces $-T''$
Deformations $-e'' \, -\Delta''$

(b)

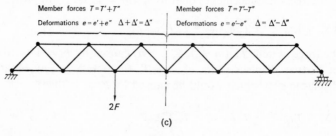

Member forces $T = T' + T''$
Deformations $e = e' + e'' \quad \Delta + \Delta' = \Delta''$

Member forces $T = T' - T''$
Deformations $e = e' - e'' \quad \Delta = \Delta' - \Delta''$

$2F$

(c)

Fig. 2.7

It may be said immediately that the forces due to that loading in corresponding members on either side of the vertical axis of symmetry such as BO and HK, are identical. Moreover, deflections of corresponding joints on either side of the axis of symmetry, such as K and O are identical. Thus, for purposes of analysis only one half of the truss need be considered. Again, if the loading is anti-symmetrical, or skew-symmetrical, as shown in Fig. 2.7(b), the forces due to that loading in corresponding members such as BO and HK are equal in magnitude but of opposite sense. Similarly, deflection of corresponding joints such as K and O are equal in magnitude but of opposite sense. Once more, for purposes of analysis only one half of the truss need be considered.

If, now, the conditions of Fig. 2.7(a) and (b) are superposed, in accordance with the principle which is valid for linear structures, then the condition shown in Fig. 2.7(c) is the result. In other words, any unsymmetrical loading of a symmetrical structure may be treated in two components, the symmetrical component and the anti-symmetrical component. Thereby analysis of half of the structure only is necessary but must be performed twice, once for the symmetrical condition and once for the anti-symmetrical condition. For each of these analyses, however, only one half of the number of equations is involved than if the unsymmetrically loaded structure is considered as a whole. Frequently, it is apparent that it is quicker to deal with one half of the overall number of equations twice than that number of equations simultaneously.

It is, incidentally, interesting to note that in the event of there being a third support of the truss shown in Fig. 2.7 located at mid-span M, then it would be unloaded due to the anti-symmetrical loading shown in Fig. 2.7(b). Therefore, it is apparent that the load on that support due to the symmetrical loading condition of Fig. 2.7(a) is the same as that due to the unsymmetrical loading condition of Fig. 2.7(c).

There are many instances in structural analysis when symmetry of loading and structural behaviour may be detected. It is important to observe these features and take every opportunity of time-saving which such features afford (see paras. 3.6, 4.3, 5.4, 5.7 and 6.8).

EXERCISES

1 Show that the horizontal deflection of joint B of the plane pin-jointed framework illustrated in Fig. 2.8 due to the loading shown is zero if the two members AB and BC are identical.

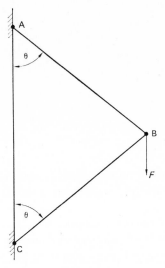

Fig. 2.8

Principles of structural analysis

2 (*a*) If the plane pin-jointed framework shown in Fig. 2.9 is such that the loading shown causes strains of the same *numerical* value of 10^{-4} in each member, calculate the deflection of joint C.

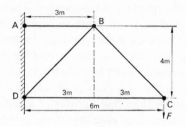

Fig. 2.9

(*b*) What is the deflection of C if a rise of temperature of the framework causes the length of each member to increase by an amount 10^{-4} times its original length?

Ans: (*a*) 2·15 mm downwards

0·60 mm to the left

downwards to the left

(*b*) 0·60 mm to the right

3 Calculate the vertical deflections of the joints B, C and D of the plane, pin-jointed framework shown in Fig. 2.10 due to the loading shown, if the flexibility of each horizontal and vertical member is *a* and that of each inclined

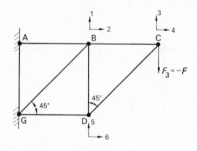

Fig. 2.10

member is $a\sqrt{2}$. Comment on the difference between the deflections of B and D and state briefly the method of calculating the horizontal deflections of B, C and D.

4 Calculate the flexibility coefficients $a_{13} = a_{31}$; $a_{23} = a_{32}$; a_{33}; $a_{43} = a_{34}$; $a_{53} = a_{35}$; $a_{63} = a_{36}$ of the framework described in problem 3 above.

5 Calculate the stiffness coefficients relating to the six degrees of freedom of the framework described in problem 3 above.

40

3

Analysis of statically indeterminate pin-jointed frameworks

3.1 As its description implies, a statically indeterminate framework cannot be analysed for the purpose of determining the forces in its members induced by loading, by means of statical principles only. It is necessary to use as well, the laws governing the elastic deformation of members caused by load and the conditions for compatibility of the strains of the framework, to obtain sufficient equations.

The provision of pin-joints is uneconomical in practice and rigid joints are, therefore, usual. For the purpose of determining axial forces in the members of frameworks the assumption of pin-joints is convenient and does not usually incur errors of more than a few per cent in spite of the presence of 'secondary effects' (see para. 6.7).

3.2 The nature of the problem: two basic approaches to analysis

For the plane framework with two degrees of freedom shown in Fig. 3.1 there are two equations of equilibrium connecting the axial forces in the members and the loads F_h and F_v. There are, however, three members one of which is redundant to the requirements of statics so that it is necessary to obtain an additional equation independent of statical considerations. This may be achieved by considering the conditions for the deflection of the loaded or free joint O to be compatible with the axial deformations of the three members which meet there, having regard to the laws of elasticity of those members.

Thus, for equilibrium of O:

$$T_1 \cos \theta_1 + T_2 \cos \theta_2 + T_3 \cos \theta_3 + F_h = 0$$
$$T_1 \sin \theta_1 + T_2 \sin \theta_2 + T_3 \sin \theta_3 + F_v = 0$$

(3.1)

while for compatibility of *small* changes in length (elastic deformation) of the

41

Principles of structural analysis

members e_1, e_2 and e_3, respectively, with the two components Δ_h and Δ_v of a *small* deflection of O:

$$e_1 = -\Delta_h \cos\theta_1 - \Delta_v \sin\theta_1$$
$$e_2 = -\Delta_h \cos\theta_2 - \Delta_v \sin\theta_2 \qquad (3.2)$$
$$e_3 = -\Delta_h \cos\theta_3 - \Delta_v \sin\theta_3$$

if extensions of members are considered to be positive.

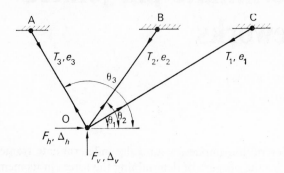

Fig. 3.1

By using the laws of elasticity of the members, e_1, e_2 and e_3 may be expressed in terms of T_1, T_2 and T_3, respectively; thus, if the members are linearly elastic:

$$e_1 = a_1 T_1; \quad e_2 = a_2 T_2; \quad e_3 = a_3 T_3 \qquad (3.3)$$

where a_1, a_2 and a_3 are the flexibilities of the members respectively. Substitution from equation (3.3) into equation (3.2) followed by elimination of Δ_h and Δ_v provides an additional equation in T_1, T_2 and T_3 which together with equations (3.1) will enable the forces in the members to be found in terms of the loads F_h and F_v. This equation represents, in fact, the condition for the axial deformation of any one of the three members as redundant to be compatible with the deformation of the statically determinate system consisting of two members. This procedure is clearly possible regardless of the nature of the elasticity, that is, whether it is linear or non-linear.

For example, if T_2 is regarded as the force in the redundant member, the member forces T_1 and T_3 may be expressed in terms of T_2, F_h and F_v by equations (3.1) and substituted in the third equation obtained in the manner indicated. If the elasticity is linear this equation will have the form:

$$(a_2 + a_{22})T_2 + a_{2h}F_h + a_{2v}F_v = 0 \qquad (3.4)$$

where a_{22}, a_{2h} and a_{2v} are flexibility coefficients of the statically determinate system consisting of members OA and OC. Thus: a_{22} is the shortening of the line OB due to unit force tending to close the gap OB in the absence of member

42

OB; a_{2h} $(= a_{h2})$ is the shortening of the line OB due to $F_h = 1$, in the absence of member OB; a_{2v} $(= a_{v2})$ is the shortening of the line OB due to $F_v = 1$ in the absence of member OB.

This procedure of analysing a statically indeterminate framework is called the *compatibility approach* or force method, because the final equation (or

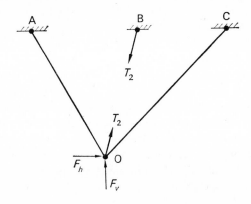

Fig. 3.2

equations, if there is more than one redundant) such as (3.4) is a strain compatibility equation with the force in the redundant as the unknown.

In order to appreciate the full physical significance of equation (3.4) it is necessary to consider the condition for the redundant strained due to a tension T_2 to fit into the statically determinate framework loaded by F_h, F_v and forces T_2, as shown in Fig. 3.2. The final length of the line or gap OB is:

$$l_2 - \Delta_2 = l_2 - a_{22}T_2 - a_{2h}F_h - a_{2v}F_v \tag{3.5}$$

while the final length of member OB is:

$$l_2 + e_2 = l_2 + a_2T_2 \tag{3.6}$$

Then for compatibility of strains, or for the fitting into place of the redundant:

$$l_2 + a_2T_2 = l_2 - a_{22}T_2 - a_{2h}F_h - a_{2v}F_v \tag{3.7}$$

from which equation (3.4) is obtained.

The alternative method of solving the problem is to use equations (3.2) and (3.3) to write:

$$a_1T_1 = -(\Delta_h \cos \theta_1 + \Delta_v \sin \theta_1)$$
$$a_2T_2 = -(\Delta_h \cos \theta_2 + \Delta_v \sin \theta_2) \tag{3.8}$$
$$a_3T_3 = -(\Delta_h \cos \theta_3 + \Delta_v \sin \theta_3)$$

and substitute the expressions, for T_1, T_2 and T_3, respectively, so obtained in equations (3.1) to obtain two equations in the loads F_h and F_v and the two unknowns Δ_h and Δ_v. In this way Δ_h and Δ_v may be found and subsequently T_1, T_2 and T_3 also, by using equations (3.8). This procedure of analysing a statically indeterminate framework is called the *equilibrium approach* or displacement method, because the final equations solved are equations of equilibrium with deflections or displacements of joints as unknowns.

If, as in this instance, the elasticity is linear the final equations obtained as described may be written:

$$F_h = b_{hh}\Delta_h + b_{hv}\Delta_v$$
$$F_v = b_{vh}\Delta_h + b_{vv}\Delta_v$$

(3.9)

where b_{hh}; $b_{hv} = b_{vh}$ and b_{vv} are stiffness coefficients of the statically indeterminate system with reference to the lines of action of F_h and F_v (see para. 2.6).

When the final equations of the two approaches are obtained directly for a linear system by the use of flexibility and stiffness coefficients, respectively, the equilibrium and compatibility conditions are implied in the calculation of these coefficients, respectively.

3.3 Choice of approach to analysis

The compatibility approach to analysis clearly yields as many final simultaneous equations relating the forces in the redundants to the loading causing those forces, as there are redundants. The equilibrium approach, on the other hand, provides final simultaneous equations relating the components of deflection of joints of the frameworks to the loading which causes them, which are as numerous as the degrees of (elastic) freedom. It follows, therefore, that comparison of the number of redundants and degrees of freedom is essential to the choice of approach. Thus, if there are appreciably more degrees of freedom than redundants, the compatibility approach is preferable for manual computation (regardless of the ease with which stiffness coefficients are calculated). This is almost always the situation regarding pinjointed frameworks and so the remainder of this chapter is largely concerned with application of the compatibility approach. It may be noted, however, that application of the equilibrium approach for linear frameworks follows precisely the contents of para. 2.6 above relating to the use of stiffness coefficients.

3.4 Application of the compatibility approach

The framework shown in Fig. 3.3 is linearly elastic, the flexibility of the inclined members being a, that of the horizontal members $4a/5$ and that of the vertical members $3a/5$. In fact is is identical to the framework considered in para. 2.7 (Fig. 2.5) above except that it contains an additional or redundant member BG. In order to analyse the framework for the forces in its members due, say, to a load F_5, as shown, it is necessary first to find the force in the redundant. Thereafter the forces in the members of the remainder of the framework can be found by using the conditions of equilibrium of the joints.

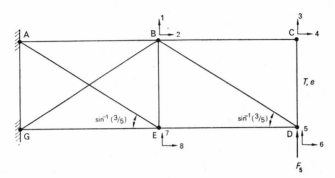

Fig. 3.3

Although several choices of redundant are possible, having regard to the framework of Fig. 2.5 member BG is the obvious choice. Assuming this member is in tension T_r due to the loading, the condition for its strain to be compatible with that of the statically determinate system is:

$$l_r + a_r T_r = l_r - a_{rr} T_r - a_{r5} F_5 \tag{3.10}$$

or

$$(a_r + a_{rr}) T_r + a_{r5} F_5 = 0 \tag{3.11}$$

where l_r is the unstrained length of both the member BG and the line BG of the statically determinate system; $a_r = a$ is the flexibility of the member BG; a_{rr} is the shortening of line BG of the statically determinate system (i.e. in the absence of member BG) due to unit forces at B and G acting toward one another in the line BG (as though there is unit tensile force in a member BG); and a_{r5} ($= a_{5r}$) is the shortening of line BG of the statically determinate system due to $F_5 = 1$.

It is noteworthy that if there are eight components of load corresponding to the eight degrees of freedom of the framework equation (3.11) becomes

$$(a_r + a_{rr}) T_r + \sum_{}^{8} a_{ri} F_i = 0 \tag{3.12}$$

Now the calculation of the flexibility coefficient $a_{r5} = (a_{5r})$ of the statically determinate system using the principle of virtual work is demonstrated in para. 2.7 above (there it is designated a_{BG5}) where its value is shown to be

45

$-2.88a$. It is then merely necessary here to calculate a_{rr}. Referring to Fig. 2.6(b), by the principle of virtual work:

$$1a_{rr} = \sum T'e' \tag{3.13}$$

the calculation for which is set out in Table 3.1.

TABLE 3.1

Member	T'	e'	$T'e'$
AB	$-4/5$	$-16a/25$	$+64a/125$
AE	$+1$	$+a$	$+a$
BE	$-3/5$	$-9a/25$	$+27a/125$
EG	$-4/5$	$-16a/25$	$+64a/125$
		$\sum =$	$+280a/125$

whence:

$$a_{rr} = 56a/25 = 2.24a \tag{3.14}$$

Substituting now for a_r, a_{rr} and a_{r5} in equation (3.11) gives:

$$T_r = 0.88F_5 \tag{3.15}$$

the positive value denoting that T_r is a tensile force in accordance with the initial assumption.

If now a second additional or redundant member CE is present in the framework as shown in Fig. 3.4 sustaining a tensile force T_s due to the load F_5, then the final equations of the compatibility approach are two-fold as follows:

$$\begin{aligned}(a_r + a_{rr})T_r + a_{rs}T_s + a_{r5}F_5 = 0 \\ a_{sr}T_r + (a_s + a_{ss})T_s + a_{s5}F_5 = 0\end{aligned} \tag{3.16}$$

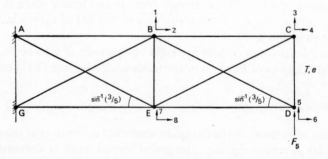

Fig. 3.4

where a_r, a_{rr} and a_{r5} have the values used above while $a_s = a$, $a_{s5} = a_{5s} = a_{(CE)5} = -2\cdot88a$ (calculated in para. 2.7), $a_{rs} = a_{sr} = a_{BGCE} = 0\cdot22a$ (calculated in para. 2.7 above) and a_{ss}, defined similarly to a_{rr}, has to be calculated.

For a_{ss}, referring to Fig. 2.6(c) and using the principle of virtual work:

$$1a_{ss} = \sum T''e'' \tag{3.17}$$

the calculation for which is set out in Table 3.2.

TABLE 3.2

Member	T''	e''	$T''e''$
BC	$-4/5$	$-16a/25$	$+64a/125$
BD	$+1$	$+a$	$+a$
BE	$-3/5$	$-9a/25$	$+27a/125$
CD	$-3/5$	$9a/25$	$+27a/125$
DE	$-4/5$	$-16a/25$	$+64a/125$
			$\sum = +307a/125$

whence:

$$a_{ss} = 307a/125 = 2\cdot456a \tag{3.18}$$

Substituting now the values of the flexibility coefficients in equations (3.16) gives:

$$3\cdot24aT_r + 0\cdot22aT_s - 2\cdot88aF_5 = 0$$
$$0\cdot22aT_r + 3\cdot46aT_s - 2\cdot88aF_5 = 0 \tag{3.19}$$

simultaneous solution of which gives:

$$T_r = 0\cdot84F_5; \qquad T_s = 0\cdot78F_5 \tag{3.20}$$

A further ('external') redundant may be introduced into the system by, say, the provision of a rigid vertical support at E with the loss of one degree of freedom, as shown in Fig. 3.5. If, due to the loading F_5 the reaction provided by the additional support is R_7, then the three-fold final equations of the compatibility approach are:

$$(a_r + a_{rr})T_r + a_{rs}T_s + a_{r7}R_7 + a_{r5}F_5 = 0$$
$$a_{sr}T_r + (a_s + a_{ss})T_s + a_{s7}R_7 + a_{s5}F_5 = 0 \tag{3.21}$$
$$a_{7r}T_r + a_{7s}T_s + a_{77}R_7 + a_{75}F_5 = 0$$

Now the additional flexibility coefficients $a_{r7} = a_{7r}$; $a_{s7} = a_{7s}$; $a_{75} = a_{57}$ and a_{77} must be calculated before these equations can be solved. As an exercise the reader is required to verify that:

$$a_{r7} = a_{7r} = -2{\cdot}54a; \qquad a_{s7} = a_{7s} = 0;$$
$$a_{75} = a_{57} = 5{\cdot}62a; \qquad a_{77} = 4{\cdot}2a \tag{3.22}$$

and by solving equations (3.21), that:

$$T_r = -0{\cdot}42F_5; \qquad T_s = 0{\cdot}86F_5; \qquad R_7 = -1{\cdot}59F_5 \tag{3.23}$$

The solution of linear simultaneous equations may be accomplished by the well-known procedure ascribed to Gauss (para. 6.6) or otherwise by means of matrix techniques (para. 9.3).

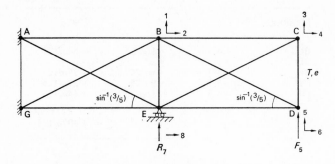

Fig. 3.5

If one or more of the redundants are absent it is merely necessary to introduce the relevant zeros. Thus, if the rigid support E is the only redundant in the structure considered above, the final equations (3.21) reduce to:

$$a_{77}R_7 + a_{75}F_5 = 0 \tag{3.24}$$

whence:

$$R_7 = -1{\cdot}34F_5 \tag{3.25}$$

It is interesting to consider here the means of taking account of the elasticity of a support. Suppose the support at E is linearly elastic instead of rigid, with a flexibility a_7. If then it exerts a reaction R_7 it follows that it must yield by an amount $-a_7R_7$, i.e. $\Delta_7 = -a_7R_7$. The minus sign is necessary because the sense of Δ_7 is opposite to that of R_7. Compatibility equation (3.24) is now modified as follows:

$$a_{77}R_7 + a_{75}F_5 = -a_7R_7 \tag{3.26}$$

or

$$(a_7 + a_{77})R_7 + a_{75}F_5 = 0 \tag{3.27}$$

If $a_7 = a$, then:

$$R_7 = -1{\cdot}08F_5 \tag{3.28}$$

3.5 Deflection of statically indeterminate frameworks

The calculation of individual components of deflection of the joints of a statically indeterminate framework due to specified loading may be achieved by considering the chosen statically determinate system subjected to the

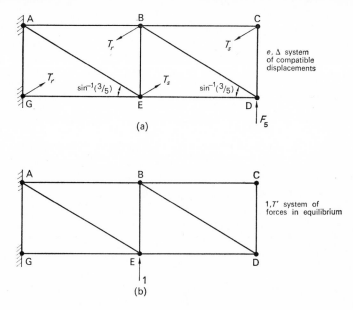

Fig. 3.6

forces in the redundants due to that loading as well as the loading itself[1]. Then if the framework is linearly elastic flexibility coefficients may be used, or otherwise, direct application of the principle of virtual work or the method of complementary energy. These latter are, of course, appropriate also for non-linear elasticity so long as the geometry of the system is not significantly changed due to the loading.

For example, to find the vertical deflection Δ_7 of joint E of the framework shown in Fig. 3.4 due to the loading F_5 shown, flexibility coefficients may be used as follows (see Fig. 3.6(a)):

$$\Delta_7 = a_{7r}T_r + a_{7s}T_s + a_{75}F_5 \tag{3.29}$$

Substituting the values of the flexibility coefficients obtained in para. 3.4 above gives:

$$\Delta_7 = 3\cdot48aF_5 \tag{3.30}$$

---

[1] This is sometimes called the "reduction method".

Alternatively, the deflection may be found by using the actual displacements associated with the force system shown in Fig. 3.6(a) as virtual displacements of the force system shown in Fig. 3.6(b) and using the principle of virtual work, thus:

$$1\Delta_7 = \sum T'e \tag{3.31}$$

i.e.

$$\Delta_7 = -(1\cdot67 \times -0\cdot83aF_5) + (1\cdot33 \times 1\cdot6aF_5) \tag{3.32}$$

Again, the method of complementary energy may be used by considering a force F_7 acting at E, as shown in Fig. 3·7. This force will clearly affect only

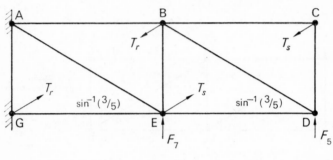

Fig. 3.7

members AE and EG (comparison with the virtual work procedure is interesting here). Having obtained the expression for the complementary energy C of the framework loaded as shown in Fig. 3.7 it will be noted that the only terms relevant to the present calculation are those for members AE and EG. If the sum of the complementary energies of these members is denoted by C', then:

$$\Delta_7 = \left(\frac{dC}{dF_7}\right)_{F_7=0} = \left(\frac{dC'}{dF_7}\right)_{F_7=0} \tag{3.33}$$

It should be noted that the calculation of the deflection of statically indeterminate frameworks is essentially the same as for statically determinate frameworks, in accordance with the contents of Chapter 2.

3.6 Example illustrating symmetry and anti-symmetry and load groups

The plane, pin-jointed framework shown in Fig. 3.8 is linearly elastic, each horizontal and vertical member having a flexibility of a and each inclined member a flexibility of $a\sqrt{2}$. The framework is thus symmetrical (elastically and geometrically) and has two redundant members. Its loading is, however, unsymmetrical as shown. In order to determine the forces in the members due to the loading it is first necessary to calculate the forces in the two redundants and it is this latter task which is the objective here. The problem is simplified if advantage is taken of the symmetry of the structure. The loading may be

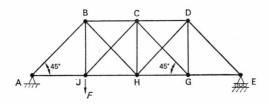

Fig. 3.8

considered as consisting of two components, a symmetrical and an anti-symmetrical component as shown in Fig. 3.9(a) and Fig. 3.9(b), respectively. The effects of these two components acting separately may be added together algebraically to give the effect of the specified load shown in Fig. 3.8. This is possible owing to the linearity of the elastic behaviour of the framework and is in accordance with the principle of superposition. Furthermore, by choosing the redundants symmetrically with respect to the vertical centre-line of the framework, for example, members CJ and CG, it is apparent that they will sustain equal forces due to the symmetrical component of the loading and forces equal in magnitude but of opposite sense, due to the anti-symmetrical component of the loading. This feature is illustrated in Fig. 3.9(a) and (b).

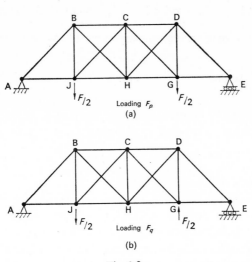

Fig. 3.9

51

Principles of structural analysis

Denoting the symmetrical type of loading shown in Fig. 3.9(a) by the symbol F_p and that shown in Fig. 3.9(b) by the symbol F_q, the following equations represent the conditions of compatibility of strain with respect to the redundant member CJ:

$$l_r - a_{rr}T_{r1} - a_{rs}T_{s1} - a_{rp}F_p = l_r + a_rT_{r1}$$
$$l_r - a_{rr}T_{r2} - a_{rs}T_{s2} - a_{rq}F_q = l_r + a_rT_{r2}$$

(3.34)

where $l_r = $ CJ in the unloaded condition; T_{r1} and T_{s1} are the forces in CJ and CG as redundants due to the loading F_p and T_{r2} and T_{s2} are the forces in them due to the loading F_q. Noting that $T_{r1} = T_{s1}$ by symmetry and $T_{r2} = -T_{s2}$ by anti-symmetry, equations (3.34) may be simplified as follows:

$$(a_r + a_{rr} + a_{rs})T_{r1} + a_{rp}F_p = 0$$
$$(a_r + a_{rr} - a_{rs})T_{r2} + a_{rq}F_q = 0$$

(3.35)

That is, instead of two simultaneous equations, two equations each in one unknown, are obtained.

Calculation of the flexibility coefficients by means of the principle of virtual work referring to Fig. 3.10 is as follows, remembering that the flexibility of horizontal and vertical members is a and of inclined members $a\sqrt{2}$:

$$a_{rr} = \sum T'e' = (4 \times a/2) + a\sqrt{2} = a\sqrt{2}(\sqrt{2} + 1)$$
$$a_{rs} = \sum T'e'' = a/2$$
$$a_{rp} = \sum T'e_p = -a/\sqrt{2} + a/\sqrt{2} = a/\sqrt{2}$$
$$a_{rq} = \sum T'e_q = a + a/2\sqrt{2} + a/\sqrt{2} = (3 + 2\sqrt{2})a/2\sqrt{2}$$

(3.36)

Substitution of these values in equations (3.35) and putting $a_r = a\sqrt{2}$ gives:

$$T_{r1} = +0{\cdot}066F; \qquad T_{r2} = +0{\cdot}238F$$

(3.37)

That is:

$$T_r = +(0{\cdot}66 + 0{\cdot}238)F = +0{\cdot}304F$$
$$T_s = -(-0{\cdot}066 + 0{\cdot}238)F = -0{\cdot}172F$$

(3.38)

where the positive sign denotes tension.

This simple example illustrates the power of the device of symmetry and anti-symmetry in the analysis of linear frameworks which are symmetrical geometrically and elastically about one axis. Thus, if there are n final equations of compatibility by this device the problem is modified so that two sets of $n/2$ simultaneous equations have to be solved.

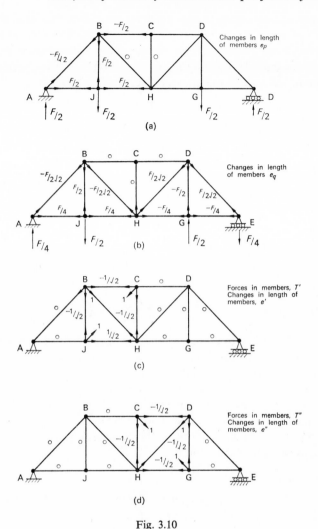

Fig. 3.10

3.7 Example of analysis of a space framework

All of the members of the freely jointed space framework shown in Fig. 3.11 are linearly elastic of the same material and cross-sectional area. It is required to calculate the force in the member OP due to the loading shown. The framework has six degrees of freedom and three redundants which indicates preference for the compatibility approach for the analysis. Moreover, examination of the problem indicates that there is no force in OP for the horizontal component of the load. This is due to the symmetrical nature of the system coupled with the symmetry of application of the horizontal component

53

of the loading, the forces from which in AO, BO, CO and DO are, by symmetry of equal magnitude, those in AO and BO being tensile while those in CO and DO are compressive. It follows also that the vertical deflection of O due to this loading is zero (Fig. 3.12) and OP is not required to provide any force to resist the horizontal loading.

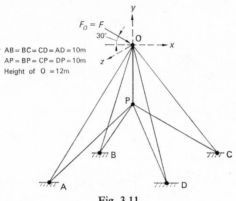

$$AB = BC = CD = AD = 10m$$
$$AP = BP = CP = DP = 10m$$
Height of O = 12m

Fig. 3.11

It remains, therefore to consider the effect of the vertical component of the load at O. Again, by symmetry, it appears that the forces in the four outer legs are equal in magnitude and that now each is in compression. There is also a compressive force in OP since vertical deflection of O occurs. Having regard to these features for the particular loading specified, the problem reduces to the determination of the force in a single redundant, namely OP.

The condition of compatibility of the strains of the redundant OP and the statically determinate system is:

$$(a_r + a_{rr})T_r + F_0 a_{r0} = 0 \qquad (3.39)$$

where T_r is the force in OP (assumed to be tensile); a_r is the flexibility of the member OP; a_{rr} the closure of the gap OP (in the absence of member OP) if unit forces at O and P act toward one another, and a_{r0} is the closure of the gap OP due to $F_0 = 1$. If, therefore, T', e' are the force and deformation systems respectively, for unit tension across the gap OP and e'' is the deformation system due to $F_0 = 1$ (without member OP), by vertical work:

$$1a_{rr} = \sum T'e' \qquad (3.40)$$

$$1a_{r0} = \sum T'e'' \qquad (3.41)$$

The member forces T' are conveniently found by resolution using O as origin; that is:

$$-4T_1' \times \frac{12}{14} - 1 = 0$$

for equilibrium of O, and   (3.42)

$$-4T'_2 \times \frac{7}{10} + 1 = 0$$

for equilibrium of P. Where T'_1 is the force (assumed to be tensile) in each of the members AO, BO, CO and DO while T'_2 is the force in each of the members AP, BP, CP and DP.

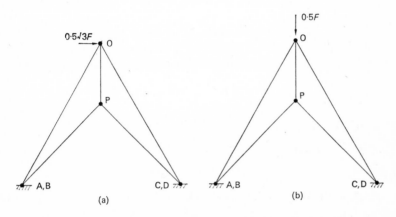

Fig. 3.12

Here the concept of *tension coefficients* is noteworthy as a useful device due to Müller-Breslau and Southwell (independently) for expressing conditions of equilibrium of space frames. It consists of writing $t = T/l$, where T is the (tensile) force in a member and l the length of that member, t being then called its tension coefficient. Equations (3.42) can be written alternatively then as follows:

$$-4t'_1 \times 12 - 1 = 0$$

$$-4t'_2 \times \ 7 + 1 = 0$$

The advantage of this concept lies in the elimination of the necessity of calculating the lengths of all of the members, all the operations being achieved in terms of the coordinates of the joints of the framework.

Thus:

$$T'_1 = -\frac{7}{24} = -0.292$$

                                                                                 (3.43)

$$T'_2 = \frac{5}{14} \ = 0.357$$

Hence if the flexibility of each of the members AP, BP, CP and DP is a, that of each of the members AO, BO, CO and DO is $1 \cdot 4a$ so that:

$$e_1' = -0 \cdot 292 \times 1 \cdot 4a = -0 \cdot 41a$$
$$e_2' = 0 \cdot 357a$$

(3.44)

Substitution in equation (3.40) then gives:

$$a_{rr} = 4([0 \cdot 292 \times 0 \cdot 41] + 0 \cdot 357^2)a$$
$$= 0 \cdot 99a$$

(3.45)

In order to calculate the e'' values, it is simply necessary to consider the vertical component of $F_0 = 1$ (namely $0 \cdot 5$) because the horizontal component of this loading causes no vertical deflection of O. Thus, for equilibrium of O:

$$-\left(4T_1'' \times \frac{12}{14}\right) - \frac{1}{2} = 0$$

(3.46)

$(T_2'' = 0)$. Whence:

$$T_1'' = -0 \cdot 146$$

(3.47)

and

$$e_1'' = -0 \cdot 146 \times 1 \cdot 4a = -0 \cdot 204a$$

(3.48)

Therefore, by equation (3.41):

$$a_{r0} = 4 \times 0 \cdot 292 \times 0 \cdot 204a$$
$$= 0 \cdot 239a$$

(3.49)

Finally, substituting for a_{rr}, a_{r0} and a_r ($= 0 \cdot 5a$) in equation (3.39) gives:

$$(0 \cdot 5 + 0 \cdot 99)T_r + 0 \cdot 239F = 0$$

(3.50)

or, the force in OP:

$$T_r = 0 \cdot 16F$$

(3.51)

By considering only the horizontal component of the loading and the four members AO, BO, CO and DO it may be shown that the deflection of O is $2 \cdot 37aF$.

3.8 Self-straining and aspects of non-linearity

Self-straining is the term used to describe forces induced in the members of a framework by factors other than applied loading. These factors include thermal effects, forcing into place of redundant members or supports during erection and settlement of foundations. They affect only statically indeterminate structures since by their very nature, statically determinate systems are immune to such factors.

As an example of self-straining, suppose the plane framework shown in Fig. 3.1 has members made of dissimilar materials and that following erection at constant temperature it suffers a temperature change such that if they were free the members would expand by small amounts λ_1, λ_2 and λ_3 respectively. In the absence of external loading the conditions of equilibrium of the free joint O are:

$$T_1 \cos \theta_1 + T_2 \cos \theta_2 + T_3 \cos \theta_3 = 0$$
$$T_1 \sin \theta_1 + T_2 \sin \theta_2 + T_3 \sin \theta_3 = 0 \tag{3.52}$$

while the conditions of compatibility of deformation are now:

$$e_1 = \lambda_1 + a_1 T_1 = -\Delta_h \cos \theta_1 - \Delta_v \sin \theta_1$$
$$e_2 = \lambda_2 + a_2 T_2 = -\Delta_h \cos \theta_2 - \Delta_v \sin \theta_2 \tag{3.53}$$
$$e_3 = \lambda_3 + a_3 T_3 = -\Delta_h \cos \theta_3 - \Delta_v \sin \theta_3$$

assuming linear elasticity, for convenience as being immaterial to the principle involved. Equations (3.53) should be compared with equations (3.8). Now either the compatibility or equilibrium approach may be adopted as described in para. 3.2 above.

In the absence of loads F_h and F_v, the components of deflection Δ_h and Δ_v of joint O are due to the thermal effect alone. If the loads are present, however, appearing in equations (3.52) as in equations (3.1), then Δ_h and Δ_v include both the thermal effect and that due to the loading.

If all the members are of the same material and the framework suffers a uniform change of temperature no self-straining occurs and Δ_h and Δ_v due to the thermal effect alone may be calculated directly using the elementary principles of thermal expansion (or contraction).

Alternatively, if the redundant member, for example, member 2, is initially too long by a small amount λ_2 so that it has to be forced into place during erection of the framework, self-straining occurs. The conditions of compatibility of deformation are now:

$$e_1 = a_1 T_1 = -\Delta_h \cos \theta_1 - \Delta_v \sin \theta_1$$
$$e_2 = \lambda_2 + a_2 T_2 = -\Delta_h \cos \theta_2 - \Delta_v \sin \theta_2 \tag{3.54}$$
$$e_3 = a_3 T_3 = -\Delta_h \cos \theta_3 - \Delta_v \sin \theta_3$$

where, if F_h and F_v are zero, Δ_h and Δ_v represent the small difference in the position of O from its position if $\lambda_2 = 0$. If, however, λ_2, F_h and F_v are not zero, Δ_h and Δ_v represent the deflection of O in relation to the position of that joint when $F_h = F_v = \lambda_2 = 0$. That is, if the deflection of O due to F_h and F_v is required the procedure is simply to put $\lambda_2 = 0$, provided the elasticity of the system is linear. If it is non-linear, however, the principle of superposition is no longer valid and it is necessary to calculate twice (including the value of λ_2), once with $F_v = F_h = 0$ and again with the load values included. In this way Δ_{ho} and Δ_{vo} are obtained, followed by Δ_h and Δ_v. The deflection caused by the loading is then represented by $\Delta_h - \Delta_{ho}$ and $\Delta_v - \Delta_{vo}$.

Confining attention now to the compatibility approach and the framework shown in Fig. 3.3 and considered in para. 3.4 above; suppose it is required to

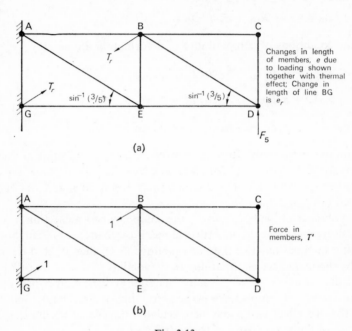

(a)

(b)

Fig. 3.13

find the force in BG as redundant due to a load F_5 together with a thermal effect such that if each member was free its length would increase by the small amount λ. Having regard to the linearity of the elasticity, this problem can be solved by treating the thermal and load effects separately. But in order to illustrate the kind of procedure which must be adopted for non-linear systems where superposition is invalid both effects will be treated simultaneously.

Calculation of the shortening (due to combined thermal and load effects) of the line BG of the statically determinate system (e_r) using the principle of virtual work is as follows:

$$1e_r = \sum T'e \tag{3.55}$$

having regard to Fig. 3.13(a) and (b). The calculation is performed in Table 3.3.

TABLE 3.3

Member*	Force T	Flexibility	Change in length e	Force T'	$T'e$
AB	$-\dfrac{4}{3}F_5 - \dfrac{4}{5}T_r$	$\dfrac{4}{5}a$	$\lambda - \dfrac{16}{15}aF_5 - \dfrac{16}{25}aT_r$	$-\dfrac{4}{5}$	$-\dfrac{4}{5}\lambda + \dfrac{64}{75}aF_5 + \dfrac{64}{125}aT_r$
AE	$-\dfrac{5}{3}F_5 + T_r$	a	$\lambda - \dfrac{5}{3}aF_5 + aT_r$	1	$\lambda - \dfrac{5}{3}aF_5 + aT_r$
BE	$F_5 - \dfrac{3}{5}T_r$	$\dfrac{3}{5}a$	$\lambda + \dfrac{3}{5}aF_5 - \dfrac{9}{25}aT_r$	$-\dfrac{3}{5}$	$-\dfrac{3}{5}\lambda - \dfrac{9}{25}aF_5 + \dfrac{27}{125}aT_r$
EG	$\dfrac{8}{3}F_5 - \dfrac{4}{5}T_r$	$\dfrac{4}{5}a$	$\lambda + \dfrac{32}{15}aF_5 - \dfrac{16}{25}aT_r$	$-\dfrac{4}{5}$	$-\dfrac{4}{5}\lambda - \dfrac{128}{75}aF_5 + \dfrac{64}{125}aT_r$

$$\sum = -1{\cdot}2\lambda - 2{\cdot}88aF_5 + 2{\cdot}24aT_r$$

* Only the four members listed enter into the calculation since $T' = 0$ for the remainder.

Now for compatibility of the strains of the statically determinate system and the redundant:

$$l_r - e_r = l_r + \lambda + a_r T_r \tag{3.56}$$

where l_r is the unstrained length of both the line BG and the redundant member BG. Substituting now from Table 3.3, $e_r = -1{\cdot}2\lambda - 2{\cdot}88aF_5 + 2{\cdot}24aT_r$ and $a_r = a$ in equation (3.56) gives:

$$3{\cdot}24aT_r = 0{\cdot}2\lambda + 2{\cdot}88aF_5 \tag{3.57}$$

or

$$T_r = 0{\cdot}062\lambda/a + 0{\cdot}889F_5 \tag{3.58}$$

The effect of self-straining is obtained simply by putting $F_5 = 0$.

Alternatively, by the method of complementary energy:

$$\delta C = \Delta_5\, \delta F_5 = \sum e\, \delta T \tag{3.59}$$

which is clearly in accordance with the principle of virtual work applied to the system of forces in equilibrium δF_5 and the δTs for compatible virtual displacements equal to those caused by the combination of the thermal and load effects. Having introduced the conditions of equilibrium such that the remaining independent variables are δF_5 and δT_r the following derivatives may be obtained:

$$\frac{\partial C}{\partial F_5} = \Delta_5$$
$$\frac{\partial C}{\partial T_r} = 0 \tag{3.60}$$

The former gives the deflection at the load due to the combined effects of the load and self-straining. The latter provides the required compatibility condition as in equation (3.57). Again, the calculation is conveniently performed in tabular form as shown in Table 3.4.

TABLE 3.4

Member*	Force T	Flexibility	Change in length e	$\dfrac{\partial T}{\partial T_r}$	$e\dfrac{\partial T}{\partial T_r}$
AB	$-\dfrac{4}{5}F_5 - \dfrac{4}{5}T_r$	$\dfrac{4}{3}a$	$\lambda - \dfrac{16}{15}aF_5 - \dfrac{16}{25}aT_r$	$-\dfrac{4}{5}$	$-\dfrac{4}{5}\lambda + \dfrac{64}{75}aF_5 + \dfrac{64}{125}aT_r$
AE	$-\dfrac{5}{3}F_5 + T_r$	a	$\lambda - \dfrac{5}{3}aF_5 + aT_r$	1	$\lambda - \dfrac{5}{3}aF_5 + aT_r$
BE	$F_5 - \dfrac{3}{5}T_r$	$\dfrac{3}{5}a$	$\lambda + \dfrac{3}{5}aF_5 - \dfrac{9}{25}aT_r$	$-\dfrac{3}{5}$	$-\dfrac{3}{5}\lambda - \dfrac{9}{25}aF_5 + \dfrac{27}{125}aT_r$
EG	$\dfrac{8}{3}F_5 - \dfrac{4}{5}T_r$	$\dfrac{4}{3}a$	$\lambda + \dfrac{32}{15}aF_5 - \dfrac{16}{25}aT_r$	$-\dfrac{4}{5}$	$-\dfrac{4}{5}\lambda - \dfrac{128}{75}aF_5 + \dfrac{64}{125}aT_r$
BG	T_r	a	$\lambda + aT_r$	1	$\lambda + aT_r$

$$\Sigma = -0{\cdot}2\lambda - 2{\cdot}88aF_5 + 3{\cdot}24aT_r$$

* Only the five members listed enter into the calculation since $\partial T/\partial T_r = 0$ for the remainder.

From Table 3.4:

$$\frac{\partial C}{\partial T_r} = -0{\cdot}2\lambda - 2{\cdot}88aF_5 + 3{\cdot}24aT_r = 0$$

or   (3.61)

$$T_r = 0{\cdot}062\lambda/a + 0{\cdot}889F_5$$

as in equation (3.58).

The virtual work and complementary energy procedures are precisely the same as described above when the elasticity is non-linear. It is merely necessary to introduce the appropriate law of elasticity for expressing that part of the change in length of members due to loading as distinct from thermal or other self-straining effect. When the elasticity is linear, however, the final equations of compatibility including self-straining effects may be expressed with the aid of flexibility coefficients. Thus, for the example considered, the single final compatibility equation, equation (3.56) may be written:

$$l_r - a_{rr}T_r - a_{r5}F_5 - \lambda_r = l_r + a_rT_r + \lambda \qquad (3.62)$$

where λ_r is the contribution to e_r due to the thermal effect. This may be calculated by applying the principle of virtual work to the system of forces in equilibrium shown in Fig. 3.13(b) for the system of compatible virtual displacements equal to those whereby each member of the statically determinate system suffers a small increase in length of λ. That is:

$$1\lambda_r = (-\tfrac{4}{5} - \tfrac{4}{5} - \tfrac{3}{5} + 1)\lambda \tag{3.63}$$

or

$$\lambda_r = -1\cdot2\lambda \tag{3.64}$$

Substituting accordingly in equation (3.62) gives, then:

$$(a_r + a_{rr})T_r + a_{r5}F_5 = 0\cdot2\lambda \tag{3.65}$$

Here, the term $0\cdot2\lambda$ represents the initial lack-of-fit of the redundant member BG due to the thermal effect. Thus, the member BG would be too short initially by $0\cdot2\lambda$ for connection into the framework following the specified thermal effect on each member.

Comparing equation (3.65) with equation (3.57) and (3.61), it appears that $a_{rr} = 2\cdot24a$ and $a_{r5} = -2\cdot88a$ which are in agreement with the values derived above in paras. 2.7 and 3.4.

EXERCISES

1 A metal truss has one redundant member. When the redundant is removed unit load applied to the structure causes the gap or line normally occupied by the redundant to extend by a units of length, while unit forces applied across the gap tending to close it causing shortening of the gap of $1\cdot6a$ units. If the flexibility of the redundant is $0\cdot15a$, calculate the force which it has to withstand when a load of 5 units is applied to the truss.

Ans: 2.85 units (tensile)

2 Calculate the horizontal thrust at the rigid abutment of the plane pin-jointed framework shown in Fig. 3.14 due to a force of 1 kN at B as shown. All members are made of the same linearly elastic material and have the same cross-sectional area with the exception of BD whose cross-sectional area is one quarter of that of the other members.

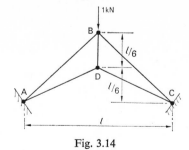

Fig. 3.14

Ans: 0·92 kN

3 In a certain framework the member AB made of light alloy ($E = 67$ kN/mm²) is redundant; it is 3·6975 m long and has a cross-sectional area of

2000 mm². When member AB is removed a tension of 10 kN between A and B in the gap in the structure (in the absence of the redundant) causes shortening of the gap by 0·25 mm. If the length of the gap AB when the structure is unstrained is 3·7000 m, calculate the force in the member AB after it is connected into the structure by force.

Ans: 48·3 kN (tensile)

4 A plane, pin-jointed framework ABCD is 1·5 m square with diagonal members AC and BD. It is connected by pins to a rigid vertical face at A and D (A being above D) and carries a load of mass 5000 kg (the force due to which is 50 kN very nearly) suspended from C. All members have a cross-sectional area of 1290 mm² and are made of steel with the exception of AC and BD which are made of light alloy. Calculate the force in the member BD if, in addition to the loading specified, the framework suffers a rise of temperature of 10°C. For steel $E = 200$ kN/mm² and the coefficient of linear expansion is 11×10^{-6}/°C; for light alloy $E = 67$ kN/mm² and the coefficient of linear expansion is 22×10^{-6}/°C.

Ans: 44 kN (compression)

5 Estimate the deflection of joint J of the plane framework with rigid joints shown in Fig. 3·15 due to the loading shown. All members are made of steel and have a cross-sectional area of 1290 mm². Comment on the accuracy of the result on the basis of assumptions made. (Consider symmetry and anti-symmetry and for calculating the deflection of J consider span KH alone having calculated first the change in length of its members.)

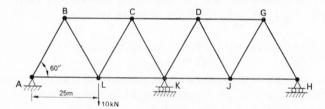

Fig. 3.15

Ans: 0·0066 mm (upward)

6 Estimate the forces in the members BN and GL and the reaction at the support at M of the linearly elastic plane framework shown in Fig. 3.16 due

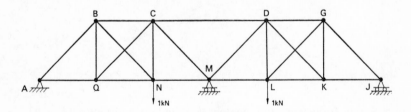

Fig. 3.16

to the specified loading. All of the members have the same cross-sectional area.

Ans: 0·32 kN (tensile); 1·5 kN (upward)

7 Find the force in the member BC of the plane, pin-jointed framework shown in Fig. 3.17 due to the loading shown. All members are linearly elastic

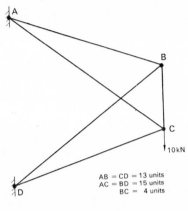

AB = CD = 13 units
AC = BD = 15 units
BC = 4 units

10kN

Fig. 3.17

with the exception of BC whose extension e is related to its tension T as follows:

$$e = 4aT(1 + 2T)$$

where a is the flexibility per unit length of each of the other members. (First find the compatibility condition for BC using virtual work or complementary energy.)

Ans: 0·44 kN (tensile)

4

Deflection, flexibility and stiffness of simple beams

4.1 It is rarely necessary nowadays to calculate deflections of linearly elastic beams from first principles having regard to the data available in both textbooks and handbooks, especially for beams of uniform section. Nevertheless, a working knowledge of principles is essential and some of the more powerful methods for deflections and flexibilities of beams are treated in this chapter where attention is confined to statically determinate systems. These methods include those of complementary energy and virtual work as well as the basic method of flexible elements, which, though considered briefly with reference to a cantilever, is of general utility. The conventional method of integration is not included, however, because it is cumbersome and is to be found already in the majority of textbooks on this subject.

The objective in most instances concerning deflection of beams should be the derivation of a few essential flexibility coefficients by means of which deflection at any point due to any condition of loading may be calculated arithmetically. In this respect it is, perhaps, surprising to find that many data for uniform beam systems in general can be derived readily from the flexibility relationships of a simple uniform cantilever.

The simple Bernoulli–Euler theory of bending of linearly elastic beams provides the well-known relationship:

$$\frac{M}{I} = \frac{\sigma}{y} = \frac{E}{R} \tag{4.1}$$

where M is the bending moment; I the second moment of area of the beam section about the neutral axis; σ the fibre stress at a distance y perpendicular from the neutral axis; E is Young's modulus of elasticity and R the radius of curvature of the beam assuming the beam was straight, i.e. $R = \infty$, before M was applied to it.

The sign convention adopted is shown in Fig. 4.1. Sagging bending moment is regarded as positive, x the distance along a beam, is positive to the right from the origin while deflection Δ of a beam is positive upward. This convention is in accordance with a right-handed Cartesian system. The alternative would be to measure x to the left when Δ downward would be positive (not the commonly used hybrid convention of x positive to the right and Δ positive downward!).

Fig. 4.1

By equation (4.1), putting $1/R = d^2\Delta/dx^2$ when $d\Delta/dx$ is small, in accordance with the assumptions of this theory of bending of beams:

$$M = EI\frac{d^2\Delta}{dx^2} \tag{4.2}$$

4.2 The flexible element approach to deflection of beams

A cantilever subjected to uniform positive bending moment is shown in Fig. 4.2. If the cantilever is rigid except for an element at P of length δx, flexural

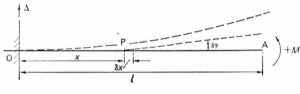

Fig. 4.2

rigidity EI and distance x from the origin O then the bending couple M causes a rotation over the element:

$$\delta\phi = M\frac{\delta x}{EI} \tag{4.3}$$

because $M/EI = 1/R$ and $\delta\phi = \delta x/R$. Also the whole of the portion PA of the cantilever rotates by $\delta\phi$ and consequently there is an upward (positive) deflection of the tip A of:

$$\delta\Delta = (l - x)\,\delta\phi = M\frac{(l - x)\,\delta x}{EI} \tag{4.4}$$

3*

It follows that if the cantilever is flexible throughout, the rotation and deflection, respectively, of A may be obtained by integrating equations (4.3) and (4.4) as follows:

$$\phi_A = M \int_0^l \frac{dx}{EI} \tag{4.5}$$

$$\Delta_A = M \int_0^l \frac{(l - x)\, dx}{EI} \tag{4.6}$$

Thus, if the cantilever is uniform such that EI is constant:

$$\phi_A = \frac{Ml}{EI} \tag{4.7}$$

$$\Delta_A = \frac{Ml^2}{2EI} \tag{4.8}$$

both of which results are well known.

In the event of the bending moment M varying along the cantilever and described by a function $M(x)$, substitution in equations (4.5) and (4.6) gives:

$$\phi_A = \int_0^l \frac{M(x)\, dx}{EI} \tag{4.9}$$

$$\Delta_A = \int_0^l \frac{M(x)(l - x)\, dx}{EI} \tag{4.10}$$

These equations are perfectly general for the slope and deflection of the tip of a cantilever for which the flexural rigidity and bending moment varies with x.

In the event of the slope and deflection of some intermediate point of the cantilever, say ϕ_x, distant X from O, being required, then it is simply necessary to replace l by X in the equations, so that, in general:

$$\phi_x = \int_0^x \frac{M(x)\, dx}{EI} \tag{4.11}$$

$$\Delta_x = \int_0^x \frac{M(x)(X - x)\, dx}{EI} \tag{4.12}$$

where $(X - x) \geqslant 0$.

Example

Calculate the slope and deflection of a uniform cantilever at a point P distant p from the root of the cantilever due to unit upward force acting at a point Q distant q from the root, as shown in Fig. 4.3.

$$M(x) = q - x$$

Therefore,

$$\phi_P = \frac{1}{EI} \int_0^p M(x)\,dx = \frac{1}{EI} \int_0^p (q - x)\,dx = \frac{p(2q - p)}{2EI} \qquad (4.13)$$

$$\Delta_P = \frac{1}{EI} \int_0^p M(x)(p - x)\,dx = \frac{1}{EI} \int_0^p (q - x)(p - x)\,dx$$

$$= p^2 \frac{(3q - p)}{6EI} \qquad (4.14)$$

The result of equation (4.14) is, in fact, the flexibility coefficient $a_{PQ} = a_{QP}$ in respect of deflection while equation (4.13) represents the flexibility coefficient $a'_{PQ} = a'_{QP}$ in respect of slope or angular deflection.

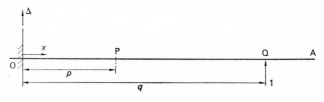

Fig. 4.3

The basic flexible element method may be used for beams generally but is then somewhat tedious and much less convenient than the alternatives derived from energy concepts. Its further application is not, therefore, considered here but the reader is recommended to apply it to a loaded simply supported beam by regarding it first as a cantilever and subsequently correcting the deflection of any point in respect of the rotation of the root of the 'cantilever', which is necessary to bring both ends of the beam to the same level.

4.3 Cantilever relationships and their application to uniform beam systems

The slope and deflection relationships for a uniform, linearly elastic cantilever due to a terminal couple, concentrated load and uniformly distributed load are given in Table 4.1 having been derived, say, by the flexible element

TABLE 4.1

Type of load on cantilever	Terminal slope	Terminal deflection
Terminal couple M	Ml/EI	$Ml^2/2EI$
Terminal force F	$Fl^2/2EI$	$Fl^3/3EI$
Uniformly distributed load of intensity w	$wl^3/6EI$	$wl^4/8EI$

method. These relationships may be used directly for any symmetrical, uniform beam system with symmetrical loading. For example, a simply supported, uniform beam, of flexural rigidity *EI*, carrying a central concentrated load, as shown in Fig. 4.4, may be regarded as two similar cantilevers springing in opposite directions from the mid-point of the beam. Each cantilever has

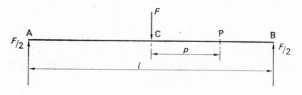

Fig. 4.4

an upward load at its tip of $F/2$, being the reactive force of the relevant support of the beam and the 'upward' deflection of the tip of each cantilever is the deflection at the centre of the beam. That is, using Table 4.1:

$$\Delta_C = -\left(\frac{F}{2}\right)\frac{(l/2)^3}{3EI} = -\frac{Fl^3}{48EI} \tag{4.15}$$

In order to find the deflection of any other point P of the beam distance, say p, from the centre it is first necessary to find the cantilever deflection Δ_P' (upward) of P and subtract from it the deflection Δ_C of the mid-point of the beam. This is illustrated in Fig. 4.5, where it is also shown that for the purpose of obtaining the cantilever deflection of P, the length p of the cantilever is

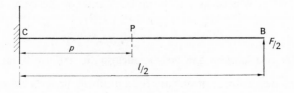

Fig. 4.5

considered subjected to an upward force $F/2$ acting at P, together with a couple $F/2(l/2 - p)$. This force and couple are clearly the external actions at the tip of the cantilever of length p which are equivalent to the internal actions at P of the cantilever when there is a load of $F/2$ at its tip. Thus, by Table 4.1 superposing the results for a cantilever with a concentrated load at its tip and then a couple there:

$$\Delta_P' = \frac{F}{2}\frac{p^3}{3EI} + \frac{F(l-2p)}{2}\frac{p^2}{2}\frac{1}{2EI} = \frac{Fp^2(3l-2p)}{24EI} \tag{4.16}$$

so that:

$$\Delta_P = \frac{Fp^2(3l - 2p)}{24EI} - \frac{Fl^3}{48EI} \tag{4.17}$$

whence:

$$a_{PC} = a_{CP} = \frac{2p^2(3l - 2p) - l^3}{48EI} \tag{4.18}$$

when $F = 1$.

The calculation of the deflection of the mid-point C of a simply supported uniform beam due to a load F acting at a distance p from that point is also readily accomplished by cantilever relationships with the aid of consideration

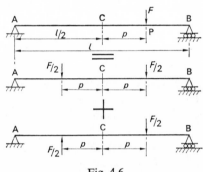

Fig. 4.6

of symmetry and anti-symmetry. For this loading may be divided into a symmetrical component and an anti-symmetrical component, as shown in Fig. 4.6. Since the latter causes no deflection of the mid-point of the beam it is merely necessary to consider the former. That is, the deflection at the centre C of a uniform beam due to a load F at any point P distant p from C is the same as the deflection there due to two separate concentrated loads $F/2$ symmetrically arranged about C at a distance of p. Now the deflection of C is the same numerically as that of a cantilever of length $l/2$ and the same flexural rigidity

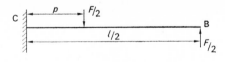

Fig. 4.7

as the beam loaded as shown in Fig. 4.7. By Table 4.1, the deflection of the tip due to a load there of $F/2$ upward is:

$$\Delta'_A = \frac{F}{2} \frac{(l/2)^3}{3EI} \tag{4.19}$$

69

and the deflection of the tip due to a load $F/2$ downward acting alone at a distance p from C is:

$$\Delta''_A = -\frac{F}{2} \times \frac{p^3}{3EI} - \frac{F}{2} \times \frac{p^2}{2EI} \times (l/2 - p) \tag{4.20}$$

Here the first term is simply the deflection of the point P of the cantilever due to a single load $F/2$ acting there while the second term is the slope at P caused by this load, multiplied by the length $(l/2 - p)$ of the unloaded portion. This

Fig. 4.8

is because the deflection of C due to this load is the same as that of P with the addition of the 'pointer effect' of the unloaded length AP, being the slope at P multiplied by $AP = (l/2 - p)$. Equation (4.17) follows upon adding (4.19) and (4.20).

Again, a couple M_A applied to the simply supported beam at A, equilibrium of the beam being maintained by equal and opposite reactions M_A/l at A and B, respectively, causes a bending condition precisely similar (Fig. 4.8) to that of a cantilever of length l with a downward load $F = -M_A/l$ at the tip. The slope at the tip of the cantilever is then:

$$\phi_B = -\frac{M_A}{l} \times \frac{l^2}{2EI} = -\frac{M_A l}{2EI} \tag{4.21}$$

while the deflection there is:

$$\Delta_B = -\frac{M_A}{l} \frac{l^3}{3EI} = -\frac{M_A l^2}{3EI} \tag{4.22}$$

so that the slope of a straight line from the root of the cantilever to its deflected tip is:

$$\psi_{AB} = -\frac{\Delta_B}{l} = -\frac{M_A l}{3EI} \tag{4.23}$$

If, therefore, the root of the cantilever is rotated anti-clockwise by ψ_{AB} to nullify the deflection of B, the beam and cantilever configurations are identical. The slope of the root of the cantilever is now:

$$\phi_A = -\psi_{AB} = \frac{M_A l}{3EI} \tag{4.24}$$

while the slope of tip is:

$$\phi_B - \psi_{AB} = -\frac{M_A l}{2EI} + \frac{M_A l}{3EI} = -\frac{M_A l}{6EI} \tag{4.25}$$

Or when $M_A = 1$:

$$\phi_A = a'_{AA} = \frac{l}{3EI}; \qquad \phi_B = a'_{BA} = a'_{AB} = -\frac{l}{6EI} \tag{4.26}$$

which are the flexibility coefficients relevant to the so-called slope–deflection relationships of uniform beams.

Example

Show that the deflection at the centre of a uniform simply supported beam of length l and flexural rigidity EI, due to a uniformly distributed load w per unit length over one half of the span is $-5wl^4/768EI$.

By considering the symmetrical and anti-symmetrical components of the loading, as shown in Fig. 4.9, it is apparent that the required deflection may

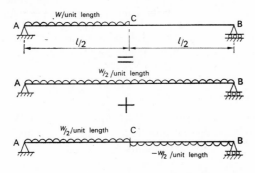

Fig. 4.9

be obtained by dealing with the beam loaded by a uniformly distributed load $w/2$ per unit length over the span.

Using now the cantilever method, the deflection at the centre of the beam is given by the deflection of the tip of a cantilever loaded as shown in Fig. 4.10:

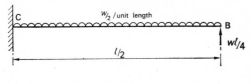

Fig. 4.10

Tip deflection due to upward load $wl/4$ there

$$\Delta_A' = \frac{(wl/4)(l/2)^3}{3EI}$$

Tip deflection due to distributed load

$$\Delta_A'' = -\frac{(wl/4)(l/2)^3}{8EI}$$

Therefore

$$\Delta_C = -(\Delta_A' + \Delta_A'') = -\frac{5(wl/4)(l/2)^3}{24EI} = \frac{-5wl^4}{768EI}$$

where the minus sign denotes downward deflection in accordance with the convention adopted.

It is interesting to note that the same result would have been obtained had the beam been loaded by a distributed load which varied linearly in intensity from w at one end of the beam to zero at the other. This is because the symmetrical component of this loading is the sole cause of deflection at the centre of the beam.

4.4 Calculation of deflections by the complementary energy method

The complementary energy (para. 1.4) of a linearly elastic beam of length l and flexural rigidity EI, in respect of bending (neglecting shearing effects) is:

$$C = \frac{1}{2}\int_0^l \frac{M^2\,dx}{EI} \tag{4.27}$$

where M, a function of x, is the bending moment caused by the loading, say, $F_1, F_2, \ldots, F_n$. By the definition of complementary energy the deflection under any one of the load components F_i is given by:

$$\Delta_i = \frac{\partial C}{\partial F_i} = \int_0^l \frac{\partial M}{\partial F_i}\frac{M\,dx}{EI} \tag{4.28}$$

For the deflection at any point i due to a load F_j applied at any other point j (or due to distributed loading), it is simply necessary to include a load F_i at i in the formulation of the bending moment M and having found $\partial M/\partial F_i$, put $F_i = 0$ throughout, that is:

$$\Delta_i = \left(\frac{\partial C}{\partial F_i}\right)_{F_i=0} = \left[\int_0^l \frac{\partial M}{\partial F_i}\frac{M\,dx}{EI}\right]_{F_i=0} \tag{4.29}$$

It should be noted here that $\Delta_i = a_{ij}F_j$ so $a_{ij} = a_{ji}$ is obtained simply by putting $F_j = 1$.

These equations may be generalised to include couples applied to a beam in which event the partial derivative of C with respect to an applied couple provides the slope of the beam at the point of application of that couple.

Example

A uniform simply supported beam AB of length l and flexural rigidity EI supports a concentrated load F_P at a point P distant p from the left-hand support at A. Find an expression for the deflection due to F_P at another point

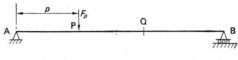

Fig. 4.11

Q distant q from that support as shown by Fig. 4.11. If there are loads F_P and F_Q at P and Q, respectively (though $F_Q = 0$, in fact), the reaction at A is:

$$R_A = \frac{l-p}{l} F_P + \frac{l-q}{l} F_Q$$

Assuming that $q > p$, the bending moment over the beam is:

between A and P:

$$M_1 = \frac{x}{l}\Big[(l-p)F_P + (l-q)F_Q\Big]$$

between P and Q:

$$M_2 = \frac{x}{l}\Big[(l-p)F_P + (l-q)F_Q\Big] - F_P(x-p)$$

between Q and B:

$$M_3 = \frac{x}{l}\Big[(l-p)F_P + (l-q)F_Q\Big] - F_P(x-p) - F_Q(x-q)$$

Now, due to the load F_P only (i.e. $F_Q = 0$):

$$\Delta_Q = \left(\frac{\partial C}{\partial F_Q}\right)_{F_Q=0} = \frac{1}{EI}\left[\int_0^p \frac{\partial M_1}{\partial F_Q} M_1 \, dx\right.$$

$$\left. + \int_p^q \frac{\partial M_2}{\partial F_Q} M_2 \, dx + \int_q^l \frac{\partial M_3}{\partial F_Q} M_3 \, dx\right]_{F_Q=0}$$

that is:

$$\Delta_Q = \frac{F_P}{EIl^2}\left\{(l-q)(l-p)\int_0^p x^2 \, dx + p(l-q)\int_p^q (l-x)x \, dx\right.$$

$$\left. + pq\int_q^l (l-x)^2 \, dx\right\}$$

73

whence

$$\Delta_Q = F_P \frac{p(l-q)}{6EIl}[2ql - (p^2 + q^2)]$$

From the solution of this example it is immediately apparent that:

$$a_{QP} = a_{PQ} = \frac{\Delta_Q}{F_P} = \frac{p(l-q)}{6lEI}[2ql - (p^2 + q^2)] \tag{4.30}$$

and when $p > q$ it is simply necessary to put $p = l - p$ and $q = l - q$.[1] The fact that $a_{QP} = a_{PQ}$ in accordance with the reciprocal theorem may be verified simply by putting $p = l - q$ and $q = l - p$. Equation (4.30) is an important result because it can be used to find the deflection of any simply supported uniform beam due to any system of concentrated loads. For example, the deflection at the centre of a uniform, simply supported beam of length l due to a uniformly distributed load of intensity w per unit length over one half of the span, i.e. from $x = 0$ to $l/2$ may be obtained as follows:

Mid-span deflection due to load $w\,\delta x$ at $p = x$ is:

$$\delta\Delta_C = \frac{x(l-l/2)}{6lEI}\left[l^2 - \left(x^2 + \frac{l^2}{4}\right)\right]w\,\delta x \tag{4.31}$$

since $q = l/2$. So that integrating for all elemental loads $w\,\delta x$ from $x = 0$ to $x = l/2$ gives:

$$\Delta_C = \frac{w}{12EI}\int_0^{l/2}\left[\frac{3l^2}{4} - x^2\right]x\,dx = \frac{5wl^4}{768EI} \tag{4.32}$$

4.5 Calculation of deflection of beams by the principle of virtual work

The essential features of the virtual work procedure are the use of virtual displacements equal to the displacements actually caused by the specified

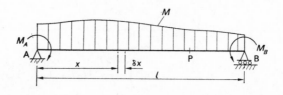

Fig. 4.12

loading of a beam, and an appropriately located arbitrary force acting on the beam with bending moments in equilibrium with it. Thus, if the bending

[1] This may be verified by imagining the beam to be viewed from behind it.

moment due to specified loads is M (Fig. 4.12) an element of the beam suffers a bending (rotational) displacement $\delta\phi = M\,\delta x/EI$. In order to find the deflection $\delta\Delta_P$ of any point P of the beam due to the effect of the loading on that single element an arbitrary concentrated load F_P' is considered to act in isolation on the beam (Fig. 4.13) so that the bending moment on the element is M', then by the principle of virtual work:

$$F_P'\,\delta\Delta_P = M'M\,\delta x/EI \tag{4.33}$$

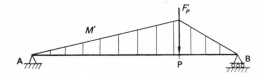

Fig. 4.13

By integration, therefore:

$$F_P'\Delta_P = \int_0^l \frac{M'M\,dx}{EI} \tag{4.34}$$

or

$$\Delta_P = \frac{1}{F_P'}\int_0^l \frac{M'M\,dx}{EI} \tag{4.35}$$

It is clearly convenient if a value of unity is chosen for F_P'.

Example

Find the deflection at the quarter span point P of a uniform simply supported beam AB of length l and flexural rigidity EI due to the bending effect of a uniformly distributed downward load of intensity w per unit length.

Bending moment due to the loading at any point distant x from A:

$$M = \frac{w}{2}(l - x)x$$

For unit load acting in isolation downward at the quarter span point P the bending moment at any point distant x from A is:

$$x \leqslant l/4, \qquad M' = \frac{3}{4}x$$

$$x \geqslant l/4, \qquad M' = \frac{3}{4}x - \left(x - \frac{l}{4}\right) = \frac{(l - x)}{4}$$

By the principle of virtual work:

$$1\Delta_P = \frac{1}{EI} \int_0^l M'M \, dx$$

$$= \frac{3w}{8EI} \int_0^{l/4} (l - x)x^2 \, dx + \frac{w}{8EI} \int_{l/4}^l (l - x)^2 x \, dx$$

$$\Delta_P = \frac{57wl^4}{6 \times 4^5 EI}$$

This is a downward deflection as denoted by the positive sign because of the choice of $F'_P = 1$ downward while the specified loading acts downward also.

The principle of virtual work affords a powerful method of calculating deflections of beams and structures generally, especially when standard formulae such as equation (4.30) are not available as, for example, for sections of variable flexural rigidity.

4.6 Use of the law of conservation of energy for finding beam deflections

The calculation of the deflection of a beam at the point of application of a single concentrated load[1] is a special and rather trivial instance in which the law of conservation of energy may be applied directly. Thus, for the deflection of, say, a uniform beam of length l and flexural rigidity EI, at a point P due to a load F_P applied, there, it is simply necessary to equate the work done, W, by the load to the strain energy, U, of the beam, as follows:

$$W = U = \int_0^{\Delta_P} F_P \, d\Delta_P = \frac{EI}{2} \int_0^l \left(\frac{d^2\Delta}{dx^2} \right)^2 dx \qquad (4.36)$$

or, putting $d^2\Delta/dx^2 = M/EI$ and noting that $\int_0^{\Delta_P} F_P \, d\Delta_P = \frac{1}{2}F_P\Delta_P$ for linear elasticity:

$$F_P\Delta_P = \frac{1}{EI} \int_0^l M^2 \, dx \qquad (4.37)$$

whence:

$$\Delta_P = \frac{1}{F_P EI} \int_0^l M^2 \, dx \qquad (4.38)$$

where M is the bending moment due to F_P. The right-hand side of equation (4.37) also represents the complementary energy of the beam for the load F_P but this is incidental, being due to the linear elasticity of the beam. Equation (4.38) may also be derived by the principle of virtual work.

Example

Find the deflection at the centre of a uniform simply supported beam of length l and flexural rigidity EI due to a concentrated downward load F applied there.

[1] Or the points of application of two such loads of equal magnitude and applied symmetrically to a symmetrical system.

By the law of conservation of energy:

$$F\Delta = \frac{1}{EI} \int_0^l M^2 \, dx$$

now when $x \leqslant l/2$:

$$M = \frac{F}{2} x$$

when $x \geqslant l/2$:

$$M = \frac{F}{2} x - F\left(x - \frac{l}{2}\right) = \frac{F}{2}(l - x)$$

so that:

$$\Delta = \frac{1}{FEI} \left[\frac{F^2}{4} \int_0^{l/2} x^2 \, dx + \frac{F^2}{4} \int_{l/2}^l (l - x)^2 \, dx \right]$$

and

$$\Delta = \frac{Fl^3}{48EI}$$

downward because in writing the work term it was assumed that Δ was of the same sense as F.

There is, however, another way in which the law of conservation of energy may be used to calculate the deflection at any point of a linearly elastic system such as a beam, due to any condition of loading. This necessitates the introduction of an arbitrary concentrated load at the point which is imagined to be applied before the actual or specified loading, so that when the latter is applied the arbitrary load does work on the structure due to the deflection whose magnitude is sought. Thus, if the point on the beam (or structure) is denoted by P and the arbitrary load is F_P', the work done on the system due to application of the specified loading $F_1, F_2, \ldots, F_n$, is:

$$W = F_P' \Delta_P + \tfrac{1}{2} \sum_{i=1}^n F_i \Delta_i \tag{4.39}$$

and the corresponding strain energy of the beam is:

$$U = \frac{1}{EI} \int_0^l M'M \, dx + \frac{1}{2EI} \int_0^l M^2 \, dx \tag{4.40}$$

where l is the span of the beam whose uniform flexural rigidity is EI, M' is the bending moment due to the arbitrary load F_P' and M is the bending moment due to the specified loading. By the law of conservation of energy $W = U$ and since by the same law applied when $F_P' = 0$:

$$\tfrac{1}{2} \sum_{i=1}^n F_i \Delta_i = \frac{1}{2EI} \int_0^l M^2 \, dx \tag{4.41}$$

it follows from equations (4.39) and (4.40) that:

$$F'_P \Delta_P = \frac{1}{EI} \int_0^l M'M \, dx \qquad (4.42)$$

or

$$\Delta_P = \frac{1}{F'_P EI} \int_0^l M'M \, dx \qquad (4.43)$$

Now this result is identical with that which is provided by the principle of virtual work though on a conceptually different basis. But while that principle is valid regardless of whether the elasticity of a structure is linear or non-linear, the same cannot be said of the method herein described. This method is conceptually correct only when the elasticity is linear and when the principle of superposition is valid, because it depends on the deflection due to specified loads being the same regardless of whether or not the structure is loaded initially (e.g. by an arbitrary load F'_P). Thus, if it were applied to a non-linear system it would provide an incorrect result which was no longer identical with that obtained by the principle of virtual work. In fact this method, which is sometimes called 'the dummy load method', is not recommended and is mentioned here purely for the sake of completeness. Even for linearly elastic systems the principle of virtual work is just as convenient if not more so.

EXERCISES

1 Find expressions for the terminal deflection of a uniform cantilever of length $2l$ and flexural rigidity EI due to a uniformly distributed load of intensity w per unit length over a distance l measured from the tip of the cantilever.

Ans: $41wl^4/24EI$

2 Find the deflection at the centre of a uniform, simply supported beam of span l and flexural rigidity EI due to a uniformly distributed load of intensity w per unit length over one-quarter of the span (measured from a support).

Ans: $23wl^4/(3 \times 8^4 EI)$

3 With the aid of the answer to problem (2) above, or otherwise, find the deflection at the centre of a uniform, simply supported beam of span l and flexural rigidity EI due to a uniformly distributed load of intensity w per unit length over three-quarters of the span.

Ans: $137wl^4/(3 \times 8^4 EI)$

4 By using an energy method, or otherwise, show that the deflection at the quarter span point of a uniform, simply supported beam of span l and flexural rigidity EI due to a concentrated load F at that point is $3Fl^3/256EI$.

5 Find the deflection at one-third span of a uniform simply supported beam of span l and flexural rigidity EI due to a uniformly distributed load of intensity w per unit length over the span.

Ans: $11wl^4/972EI$

5

Analysis by the compatibility approach of simple statically indeterminate beams, arches and portal frameworks

5.1 While the compatibility approach (para. 3.2) is not usually preferable for the analysis of complex beam systems and rigidly jointed frameworks, it is useful for dealing with some of the more elementary systems of these kinds and is essential for the analysis of arches.

The contents of this chapter (in common with those of Chapter 4) have as their basis the elementary theory of bending of structural elements, which includes the assumption of linear elasticity. Axial and shear deformations are neglected since they are usually of little significance in practice. Moreover, attention is confined to beams of uniform section as being common in engineering practice and sufficient for demonstrating principles. The treatment of arches is, however, general in respect of flexure.

5.2 Statically indeterminate beam systems having one redundant

The propped cantilever is one of the simplest statically indeterminate beam systems. Thus, in order to calculate the reaction exerted by a rigid prop (as the chosen redundant[1]) at the end of a uniform cantilever of length l and flexural rigidity EI subjected to a downward uniformly distributed load of intensity w per unit length as shown in Fig. 5.1, it is necessary to establish the relevant condition of compatibility of deformation. That is, the condition for the deflection of the cantilever to be zero at the prop. Referring to Table 4.1, in

---

[1] An alternative choice of redundant is the restraining couple at the root of the cantilever. This kind of choice of redundant is not considered herein for any beam system because it is not advantageous or meritorious instructionally.

the absence of the prop the downward deflection of the tip of the cantilever due to the specified loading is:

$$\Delta_B' = -\frac{wl^4}{8EI} \tag{5.1}$$

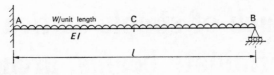

Fig. 5.1

Due to an upward force R_B acting at the tip of the cantilever, however, the upward deflection there is:

$$\Delta_B'' = a_{BB}R_B = \frac{l^3}{3EI}R_B \tag{5.2}$$

Now for zero deflection of **B**:

$$\Delta_B = \Delta_B' + \Delta_B'' = 0 \tag{5.3}$$

that is,

$$-\frac{wl^4}{8EI} + \frac{l^3}{3EI}R_B = 0 \tag{5.4}$$

or

$$R_B = \frac{3wl}{8} \tag{5.5}$$

If, however, the prop is linearly elastic such that it sinks by an amount a_B per unit reaction which it exerts, then the deflection at B is no longer zero, being in fact a_BR_B downward. The compatibility equation thus becomes:

$$\Delta_B = \Delta_B' + \Delta_B'' = -a_BR_B \tag{5.6}$$

or

$$-\frac{wl^4}{8EI} + \frac{l^3}{3EI}R_B = -a_BR_B \tag{5.7}$$

whence:

$$R_B = \frac{3wl^4}{8(3EIa_B + l^3)} \tag{5.8}$$

Again, if the load on the cantilever is a concentrated load F at mid-span:

$$\Delta_B' = -\frac{(l/2)^3}{3EI}F - \frac{(l/2)^3}{2EI}F \tag{5.9}$$

where the two terms represent the deflection at the point of application of the load and the slope there multiplied by the distance from the load to the tip of the cantilever (Table 4.1). Thus, the latter takes account of the deflection of the tip of the cantilever due to the straight unloaded part of its length rotating through the small angle of slope at the load, caused by bending of the loaded part.

Substituting now from equation (5.9) in equation (5.3) assuming a rigid prop, gives:

$$-\frac{5l^3}{48EI} F + \frac{l^3}{3EI} R_B = 0 \tag{5.10}$$

or

$$R_B = \frac{5}{16} F \tag{5.11}$$

If the prop is elastic it is merely necessary to introduce the term $a_B R_B$ as in equation (5.7).

Suppose now the prop is at C half-way along the cantilever and that the concentrated load is at the tip. Then:

$$\Delta_C' = -\frac{(l/2)^3}{3EI} F - \frac{(l/2)^3}{2EI} F \tag{5.12}$$

Where, with reference to Table 4.1, the first term is the deflection of C due to a concentrated load F acting there and the second term is the deflection of C due to the action of a clockwise couple $Fl/2$ there.[1] The force and couple are, in fact, the equivalent actions at C of the concentrated load at the tip of the cantilever. Now:

$$\Delta_C'' = \frac{(l/2)^3}{3EI} R_C \tag{5.13}$$

and for zero deflection of C:

$$\Delta_C = \Delta_C' + \Delta_C'' = 0 \tag{5.14}$$

or

$$\frac{l^3}{24EI} R_C = \frac{5l^3}{48EI} F \tag{5.15}$$

whence:

$$R_C = \frac{5}{2} F \tag{5.16}$$

[1] In the event of the load on the overhanging part of the cantilever being uniformly distributed of intensity w per unit length the effective load force and couple at C are $-wl/2$ and $-wl^2/4$, respectively.

Having calculated the force exerted by the redundant in all of the problems considered herein, the elements of statics are sufficient for the purpose of calculating the forces and couples acting throughout each system. Deflections may then be found, if necessary, using methods such as those described in Chapter 4.

Another simple statically indeterminate beam system is shown in Fig. 5.2

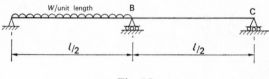

Fig. 5.2

where the beam AC is uniform of flexural rigidity *EI*. Consideration of the intermediate support mid-way between A and C affords some interesting applications of symmetry and anti-symmetry while at the same time illustrating general principles. Thus, if there is a uniformly distributed load of intensity *w* per unit length over the span AB calculation of the force exerted by the centre support, as the redundant, is greatly facilitated by making use of symmetry and anti-symmetry and superposition as in para. 4.3, Chapter 4. Reference to Fig. 5.3 shows that in the absence of the support at B deflection

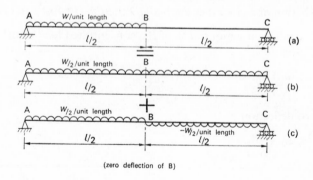

(zero deflection of B)

Fig. 5.3

there due to the specified loading is the same as though there is a uniformly distributed load of intensity *w*/2 per unit length extending from A to C. In order to find the reaction exerted by the support at B it is, therefore, necessary to consider the symmetrical component of the loading only.

Similarly, if there is a distributed load varying linearly in intensity from zero at A to *w* at C, the reaction at support B may be found by considering a uniformly distributed load of intensity *w*/2 per unit length extending from A to C, as shown in Fig. 5.4. Moreover, the result will be the same as for the loading shown in Fig. 5.2.

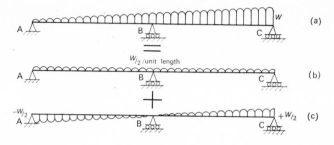

Fig. 5.4

Proceeding then and considering the loading shown in Fig. 5.3(b) and Fig. 5.4(b) for the purpose of finding R_B, the relevant compatibility condition specifies zero deflection at B. Thus:

$$\Delta_B = \Delta_B' + \Delta_B'' = 0 \tag{5.17}$$

In para. 4.3 it is shown that:

$$\Delta_B' = -\frac{5wl^4}{768EI} \tag{5.18}$$

while equation (4.15) indicates that:

$$a_{BB} = \frac{l^3}{48EI} \tag{5.19}$$

therefore:

$$\Delta_B'' = a_{BB}R_B = \frac{l^3}{48EI} R_B \tag{5.20}$$

Substitution from equations (5.18) and (5.20) in equation (5.17) then gives:

$$-\frac{5wl^4}{768EI} + \frac{l^3}{48EI} R_B = 0 \tag{5.21}$$

so that:

$$R_B = \frac{5wl}{16} \tag{5.22}$$

Other configurations of a beam with intermediate support and various conditions of loading may be treated by using the general expression for beam flexibility coefficients of equation (4.30) for the purpose of calculating both Δ_B' and Δ_B''. The method of calculating the reaction exerted by an elastic support is precisely the same as that used above for the propped cantilever.

5.3 Statically indeterminate beam systems having two redundants

The uniform encastré beam shown in Fig. 5.5 is an example of a statically indeterminate beam system with two redundants. Thus, the restraints (couples) against slope of the ends of the beam may be chosen as the redundants or

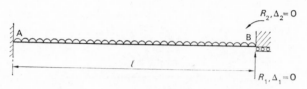

Fig. 5.5

alternatively the two-fold restraints at one end of the beam. Choice of the latter enables the relationships in respect of cantilever behaviour given in Table 4.1 to be used for finding the redundant force and couple due to specified loading. The compatibility conditions which enable the redundants to be resolved specify zero deflection and slope of B. Denoting these quantities by Δ_1 and Δ_2, respectively, the deflection and slope of B caused by the specified loading in the absence of restraint at B, by Δ_1' and Δ_2' respectively, and the deflection and slope of B due to the reactive effects there by Δ_1'' and Δ_2'', then for compatibility of deformation:

$$\Delta_1 = \Delta_1' + \Delta_1'' = 0$$
$$\Delta_2 = \Delta_2' + \Delta_2'' = 0 \tag{5.23}$$

If the beam carries a uniformly distributed load of intensity w per unit length over the whole of its span:

$$\Delta_1' = -\frac{wl^4}{8EI}$$

$$\Delta_2' = -\frac{wl^3}{6EI} \tag{5.24}$$

as shown in Table 4.1. While:

$$\Delta_1'' = a_{11}R_1 + a_{12}R_2$$
$$\Delta_2'' = a_{21}R_1 + a_{22}R_2 \tag{5.25}$$

where, by Table 4.1:

$$a_{11} = \frac{l^3}{3EI}; \qquad a_{12} = \frac{l^2}{2EI}$$

$$a_{21} = \frac{l^2}{2EI}; \qquad a_{22} = \frac{l}{EI} \tag{5.26}$$

Substituting accordingly in equations (5.23) gives finally:

$$\Delta_1 = \Delta_1' + a_{11}R_1 + a_{12}R_2 = -\frac{wl^4}{8EI} + \frac{l^3}{3EI}R_1 + \frac{l^2}{2EI}R_2 = 0$$

$$\Delta_2 = \Delta_2' + a_{21}R_1 + a_{22}R_2 = -\frac{wl^3}{6EI} + \frac{l^2}{2EI}R_1 + \frac{l}{EI}R_2 = 0$$

(5.27)

However, by symmetry it is apparent that $R_1 = wl/2$ so that only one of these equations is necessary to show that $R_2 = wl^2/12$.

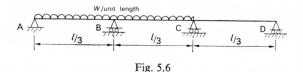

Fig. 5.6

A further example of a beam system with two redundants is afforded by the continuous beam shown in Fig. 5.6. Treating the intermediate rigid supports at B and C as redundants exerting reactions R_B and R_C, respectively, due to the loading shown, the final equations of compatibility for the determination of R_B and R_C are as follows:

$$\Delta_B = \Delta_B' + \Delta_B'' = 0$$

$$\Delta_C = \Delta_C' + \Delta_C'' = 0$$

(5.28)

where:

$$\Delta_B' = -w \int_0^{2l/3} a_{BX}\, dx$$

$$\Delta_C' = -w \int_0^{2l/3} a_{CX}\, dx$$

(5.29)

being the deflections at B and C, respectively, in the absence of supports at these points, and:

$$\Delta_B'' = a_{BB}R_B + a_{BC}R_C$$

$$\Delta_C'' = a_{CB}R_B + a_{CC}R_C$$

(5.30)

The flexibility coefficients a_{BX}, a_{CX}, a_{BB}, a_{CC} and $a_{BC} = a_{CB}$ may be calculated by means of equation (4.30). It should be noted that a_{BX} and a_{CX} represent the deflections at B and C, respectively, due to unit force acting at any point X distant x from A. Thus, the deflections at these points due to an element of distributed load, $w\, \delta x$, at X are $-a_{BX}w\, \delta x$ and $-a_{CX}w\, \delta x$, respectively. Summation of these effects gives Δ_B' and Δ_C' for the specified loading in accordance with equations (5.29). Applying equation (4.30) gives, then:

$$a_{BX} = x[2l^2/3 - (x^2 + l^2/9)]/9EI$$

(5.31)

for the value of x from 0 to $l/3$;

$$a_{BX} = (l - x)\{4l^2/3 - [(l - x)^2 + 4l^2/9]\}/18EI \qquad (5.32)$$

for values of x from $l/3$ to $2l/3$, i.e. putting $p = (l - x)$ and $q = l - l/3$ in equation (4.30);

$$a_{CX} = x[4l^2/3 - (x^2 + 4l^2/9)]/18EI \qquad (5.33)$$

$$a_{BB} = l[2l^2/3 - (l^2/9 + l^2/9)]/27EI = 8l^3/486EI \qquad (5.34)$$

which by symmetry is equal to a_{CC};

$$a_{CB} = a_{BC} = l[4l^2/3 - (l^2/9 + 4l^2/9)]/54EI = 7l^3/486EI \qquad (5.35)$$

Using the expression for a_{BX} and a_{CX} of equations (5.31), (5.32) and (5.33) in equation (5.29) gives:

$$\Delta'_B = -17wl^4/1944EI; \qquad \Delta'_C = -16wl^4/1944EI \qquad (5.36)$$

Finally, therefore:

$$\begin{aligned}
\Delta_B &= \Delta'_B + a_{BB}R_B + a_{BC}R_C \\
&= (-17wl + 32R_B + 28R_C)l^3/1944EI = 0 \\
\Delta_C &= \Delta'_C + a_{CB}R_B + a_{CC}R_C \\
&= (-16wl + 28R_B + 32R_C)l^3/1944EI = 0
\end{aligned} \qquad (5.37)$$

whence:

$$R_B = 0·40wl; \qquad R_C = 0·15wl \qquad (5.38)$$

Alternatively, the loading may be considered in terms of three components, two symmetrical and one anti-symmetrical, as shown in Fig. 5.7. By this means three independent equations each having one unknown representing

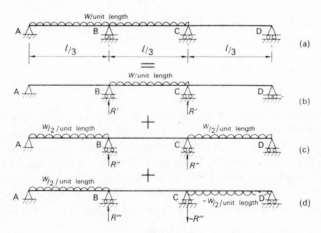

Fig. 5.7

a component of the reactions exerted by the supports at B and C may be obtained. Calculation of the various values of Δ'_B and Δ'_C is, however, laborious.

It is a simple matter to deal with elasticity of intermediate supports since the only change in equations (5.37) is the addition of terms $a_B R_B$ and $a_C R_C$ as follows:

$$\Delta_B = \Delta'_B + a_{BB}R_B + a_{BC}R_C = -a_B R_B$$

$$\Delta_C = \Delta'_C + a_{CB}R_B + a_{CC}R_C = -a_C R_C \tag{5.39}$$

or

$$\Delta'_B + (a_B + a_{BB})R_B + a_{BC}R_C = 0$$

$$\Delta'_C + a_{CB}R_B + (a_C + a_{CC})R_C = 0 \tag{5.40}$$

Problems involving more than two redundants may be solved in the manner described above. The final equations of compatibility are as numerous as the redundants, and may be solved in the manner indicated in para. 6.6 or by matrix or relaxation techniques (paras. 9.3 and 9.5).

In the event of one or both ends of a continuous beam being encastré, the requisite flexibility coefficients may be deduced by means of the data given in Table 4.1. (Equation (4.30) is appropriate only when the extremities of the beam are simply supported.)

The type of loading considered herein is generally representative of what is usually called live loading. In addition, there is always dead loading on beam systems, including the self-weight of the beam. Indeed, one of the main reasons for introducing statically indeterminate support systems is that of the self-weight of the structure.

5.4 Analysis of statically indeterminate elastic arches
The unsymmetrical linearly elastic arch shown in Fig. 5.8 is encastré in the abutments at O and Q. It has three redundants since removal of one abutment

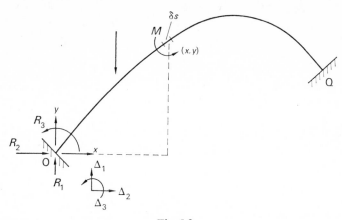

Fig. 5.8

with its three components of reaction to deformation of the arch would not destroy the system as a structure capable of carrying load. Choosing the quantities representing the redundants as R_1, R_2 and R_3 at O as shown, the bending moment M at any element of length δs measured along the arch, whose location is defined by the coordinates x, y with reference to the origin at O is:

$$M = M_0 - R_1 x + R_2 y + R_3 \tag{5.41}$$

where M_0 is the statically determinate bending moment at (x, y) due to the loading in the absence of the abutment at O. Due to this bending moment the element will bend through an angle:

$$\delta\phi = \frac{M\,\delta s}{EI} \tag{5.42}$$

by the elementary theory of bending considered in Chapter 4, where EI is the flexural rigidity. This bending will cause deflection of O (supposing the abutment is elastic) of:

$$\delta\Delta_1 = -x\,\delta\phi = -\frac{Mx\,\delta s}{EI}; \qquad \delta\Delta_2 = y\,\delta\phi = \frac{My\,\delta s}{EI};$$

$$\delta\Delta_3 = \delta\phi = \frac{M\,\delta s}{EI} \tag{5.43}$$

assuming small deformation of the arch such that there is no significant change in coordinates (x, y) following application of load. Substituting now for M from equation (5.41) and integrating to include the behaviour of all of the elements of which the arch is made, gives the final equations of compatibility of deformation at O. Thus, if the abutment at O is rigid:

$$\Delta_1 = -\int_0^Q \frac{M_0 x\,ds}{EI} + R_1\int_0^Q \frac{x^2\,ds}{EI} - R_2\int_0^Q \frac{xy\,ds}{EI} - R_3\int_0^Q \frac{x\,ds}{EI} = 0$$

$$\Delta_2 = \int_0^Q \frac{M_0 y\,ds}{EI} - R_1\int_0^Q \frac{xy\,ds}{EI} + R_2\int_0^Q \frac{y^2\,ds}{EI} + R_3\int_0^Q \frac{y\,ds}{EI} = 0$$

$$\Delta_3 = \int_0^Q \frac{M_0\,ds}{EI} - R_1\int_0^Q \frac{x\,ds}{EI} + R_2\int_0^Q \frac{y\,ds}{EI} + R_3\int_0^Q \frac{ds}{EI} = 0 \tag{5.44}$$

or

$$\Delta_1 = \Delta_1' + a_{11}R_1 + a_{12}R_2 + a_{13}R_3 = 0$$
$$\Delta_2 = \Delta_2' + a_{21}R_1 + a_{22}R_2 + a_{23}R_3 = 0 \tag{5.45}$$
$$\Delta_3 = \Delta_3' + a_{31}R_1 + a_{32}R_2 + a_{33}R_3 = 0$$

where the expressions for the flexibility coefficients with their sign may be identified in equations (5.44). These equations are sufficient to enable R_1, R_2 and R_3 to be calculated for prescribed loading which determines M_0.

The flexibility coefficients may be derived alternatively by means of the principle of virtual work by considering the force and deformation systems shown in Fig. 5.9. Thus, for example:

$$1a_{11} = \int_0^Q M' \, d\phi'; \qquad 1a_{12} = \int_0^Q M' \, d\phi'';$$

$$1a_{13} = \int_0^Q M' d\phi''' \tag{5.46}$$

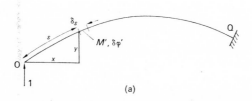

(a)

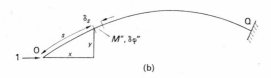

(b)

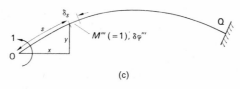

(c)

Fig. 5.9

where:

$$\delta\phi' = \frac{M' \, \delta s}{EI}; \qquad \delta\phi'' = \frac{M'' \, \delta s}{EI}; \qquad \delta\phi''' = \frac{M''' \, \delta s}{EI}. \tag{5.47}$$

In the event of the arch being pin-jointed to the abutment at O, $R_3 = 0$, and the final equations of compatibility are two-fold:

$$\Delta_1 = \Delta'_1 + a_{11}R_1 + a_{12}R_2 = 0$$
$$\Delta_2 = \Delta'_2 + a_{21}R_1 + a_{22}R_2 = 0 \tag{5.48}$$

where the values of the flexibility coefficients are as in equations (5.44). Again,

4+

Principles of structural analysis

if the arch is pin-jointed at both abutments only R_2 is statically indeterminate and the single final equation of compatibility is:

$$\Delta_2 = \Delta_2' + a_{22}R_2 = 0 \tag{5.49}$$

The relevant statically determinant system is shown in Fig. 5.10.

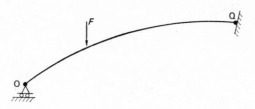

Fig. 5.10

Fortunately it is usually desirable to design symmetrical arches in engineering practice. Then further simplification is afforded. Suppose a symmetrical arch encastré in rigid abutments is subjected to the loading shown in Fig. 5.11(a). The problem may be considered in two parts as shown in Fig. 5.11(b) and (c) by the use of symmetry and anti-symmetry. Choosing the origin at the

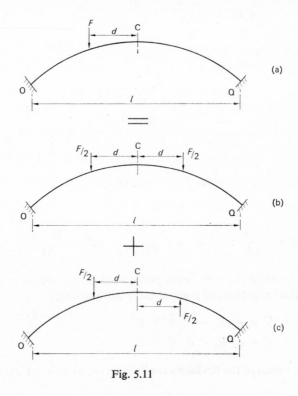

Fig. 5.11

crown C, as shown in Fig. 5.12, for the symmetrical component of the loading it follows by symmetry that both the horizontal and rotational deflections of C, Δ_2 and Δ_3, respectively, are zero; also that R_1 is zero. R_1 is zero because

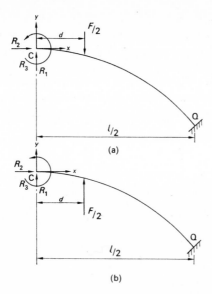

Fig. 5.12

by symmetry there is a vertical reaction at Q of $F/2$. Thus, with reference to the origin at C:

$$\Delta_2 = \Delta_2' + a_{22}R_2 + a_{23}R_3 = 0$$
$$\Delta_3 = \Delta_3' + a_{32}R_2 + a_{33}R_3 = 0 \tag{5.50}$$

where

$$a_{22} = \int_C^Q \frac{y^2 \, ds}{EI}; \qquad a_{23} = a_{32} = \int_C^Q \frac{y \, ds}{EI}; \qquad a_{33} = \int_C^Q \frac{ds}{EI}$$
$$\Delta_2' = \int_C^Q \frac{M_0 y \, ds}{EI}; \qquad \Delta_3' = \int_C^Q \frac{M_0 \, ds}{EI} \tag{5.51}$$

For the anti-symmetrical component of the loading there is a point of contra-flexure at C with in consequence zero bending moment (i.e. $R_3 = 0$) and vertical deflection (i.e. $\Delta_1 = 0$). Moreover, the horizontal abutment thrusts caused by the downward load $F/2$ on one half of the arch are nullified by the equal and opposite thrusts induced by the upward load $F/2$ on the other half of the arch. That is $R_2 = 0$ at C also. There is, then, only one unknown to be found, namely R_1, the relevant compatibility equation being:

$$\Delta_1 = \Delta_1' + a_{11}R_1 = 0 \tag{5.52}$$

where $\Delta_1' = -\int_C^Q M_0 x \, ds/EI$.

The solution of the original problem is provided by combining the results of the two solutions. It is interesting to note that the symmetrical component of the loading is responsible alone for R_2 and R_3.

Example

It is required to find the bending moment at the crown of a uniform encastré arch in the shape of a circular arc, of span 100 m and rise 25 m, due to a uniformly distributed load of 10^3 kg $= 10$ kN per horizontal m over the span.

Having regard to the symmetry of the system it is desirable to consider one half of the arch only and to adopt the crown C as the origin for the system of coordinates, as shown in Fig. 5.13. Of the three redundants at C, $R_1 = 0$ by

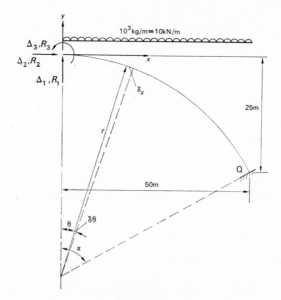

Fig. 5.13

symmetry and R_2 and R_3 are unknown and may be determined by the final equations of compatibility of deformation:

$$\Delta_2 = 0 = \Delta_2' + a_{22}R_2 + a_{23}R_3$$

$$\Delta_3 = 0 = \Delta_3' + a_{32}R_2 + a_{33}R_3$$

where:

$$a_{22} = \int_C^Q \frac{y^2\,\mathrm{d}s}{EI}; \qquad a_{23} = a_{32} = \int_C^Q \frac{y\,\mathrm{d}s}{EI}; \qquad a_{33} = \int_C^Q \frac{\mathrm{d}s}{EI};$$

$$\Delta_2' = \int_C^Q \frac{M_0 y\,\mathrm{d}s}{EI}; \qquad \Delta_3' = \int_C^Q \frac{M_0\,\mathrm{d}s}{EI}$$

Now, with reference to Fig. 5.13:

$$50^2 = 25(2r - 25)$$

by the properties of a circle, so that:

$$r = 62 \cdot 5 \text{ m}$$

and the angle subtended by the arc CQ at the centre of the circle of which the arch is part is:

$$a = \sin^{-1} 50/62 \cdot 5 = 0 \cdot 927 \text{ rad}$$

Also:

$$x = r \sin \theta = 62 \cdot 5 \sin \theta$$
$$y = r (\cos \theta - 1) = 62 \cdot 5 (\cos \theta - 1)$$
$$\delta s = r \delta \theta = 62 \cdot 5 \, \delta \theta$$
$$M_0 = 10x^2/2 = (10 \times 62 \cdot 5^2 \sin^2 \theta)/2$$

Therefore:

$$a_{22} = \frac{62 \cdot 5^3}{EI} \int_0^{0 \cdot 927} (\cos \theta - 1)^2 \, d\theta = \frac{7556}{EI}$$

$$a_{23} = a_{32} = \frac{62 \cdot 5^2}{EI} \int_0^{0 \cdot 927} (\cos \theta - 1) \, d\theta = -\frac{497}{EI}$$

$$a_{33} = \frac{62 \cdot 5}{EI} \int_0^{0 \cdot 927} d\theta = \frac{58}{EI}$$

$$\Delta_2' = -\frac{5 \times 62 \cdot 5^4}{EI} \int_0^{0 \cdot 927} \sin^2 \theta (1 - \cos \theta) \, d\theta = -\frac{40 \cdot 43}{EI} \times 10^5$$

$$\Delta_3' = \frac{5 \times 62 \cdot 5^3}{EI} \int_0^{0 \cdot 927} \sin^2 \theta \, d\theta = \frac{2 \cdot 73}{EI} \times 10^5$$

Finally then:

$$-40 \cdot 43 \times 10^5 + 7556R_2 - 497R_3 = 0$$
$$2 \cdot 73 \times 10^5 - 497R_2 + 58R_3 = 0$$

whence:

$$R_2 = 517 \text{ kN}; \qquad R_3 = -270 \text{ kN m (i.e. clockwise)}$$

The bending moment at the crown is therefore 270 kN m sagging.

Alternatively the coefficients may be calculated numerically as follows:

$$a_{22} = \frac{\delta s}{EI} \sum_{}^{n} y_i^2$$

$$a_{23} = a_{32} = \frac{\delta s}{EI} \sum_{}^{n} y_i$$

$$a_{33} = \frac{n \, \delta s}{EI}$$

where δs is the length of n equal parts into which the arc CQ is divided and y_i is the y coordinate of the ith such part.

Similarly:

$$\Delta_2' = \frac{5 \times \delta s}{EI} \sum_{}^{n} x_i^2 y_i$$

$$\Delta_3' = \frac{5 \times \delta s}{EI} \sum_{}^{n} x_i^2$$

If $n = 5$, $\delta s = 11 \cdot 6$ m and the coordinates and their functions are shown in the table below:

i	x	y	x^2	y^2	$x^2 y$
1	5·9	−0·4	35	0·1	13
2	17·1	−2·4	292	5·9	713
3	27·9	−6·6	778	44·1	5137
4	37·7	−12·6	1421	158·8	17 908
5	46·1	−20·4	2125	416·2	43 345
$\sum$		−42·5	4651	625·1	67 125
$\delta s \sum$		−493·0	5396	7251·0	778 650

whence:

$$a_{22} = 7251/EI; \qquad a_{23} = a_{32} = -493/EI; \qquad a_{33} = 58/EI;$$
$$\Delta_2' = -38 \cdot 93 \times 10^5/EI; \qquad \Delta_3' = 2 \cdot 698 \times 10^5/EI$$

It is noteworthy that it is frequently sufficient to use a relatively small number of elements to obtain results which are satisfactory for practical purposes.

5.5 Thrust line of an arch

The thrust line of an arch encastré in rigid abutments subjected to a concentrated load is shown in Fig. 5.14. It represents the direction of the resultant

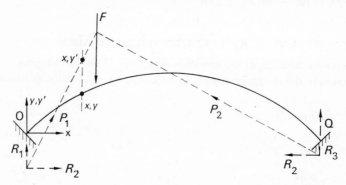

Fig. 5.14

force acting on the arch at any point, the horizontal component of which is the horizontal component R_2 of the abutment thrust. If the coordinates of a point of the thrust line are (x, y') then the bending couple acting on the arch at point (x, y) is:

$$M = R_2(y - y') \tag{5.53}$$

If the arch is pin-jointed to the abutments then it follows that the thrust line passes through the pin-joints.

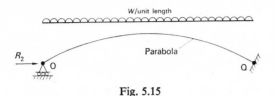

Fig. 5.15

The thrust line due to a load distributed over the whole span of an arch is a continuous curve. If the arch encastré in rigid abutments is symmetrical and parabolic in shape and if the load is uniformly distributed, the thrust line is coincident with the line of the arch and the bending moment everywhere is zero. For this condition of loading there is no difference, then, between a parabolic arch encastré in rigid abutments and one which is pin-jointed to rigid abutments. For justification of these statements, consider a uniformly loaded statically determinate parabolic arch as shown in Fig. 5.15 pin-jointed to a rigid abutment at Q and rigidly supported at O with freedom for rotation and horizontal deflection. The bending moment distribution $M(x)$ is parabolic as for a simply supported beam subjected to a uniformly distributed load over the whole of its span. If now a force R_2 is introduced at O, sufficient to nullify the horizontal deflection caused by the load, it will also cause a parabolic bending moment (due to the parabolic shape of the arch) of opposite sense to that caused by the load. It follows then that the deflection of O will be restored to zero when R_2 is of such magnitude that the bending moment in the arch caused by the load is just equal to that caused by R_2. For this value of R_2, then, the thrust line is coincident with the line of the arch.

5.6 The concept of the elastic centre

By regarding $\delta s/EI$ as an element of 'elastic area' A, the flexibility coefficients of equations (5.44) may be expressed in terms of the elastic area A, the first moments of elastic area, $A\bar{x}$ and $A\bar{y}$ and the second moments of elastic area, I_{xx}, I_{yy} and I_{xy}, thus:

$$\begin{aligned}
\Delta_1 &= \Delta_1' + I_{yy}R_1 - I_{xy}R_2 - A\bar{x}R_3 = 0 \\
\Delta_2 &= \Delta_2' - I_{xy}R_1 + I_{xx}R_2 + A\bar{y}R_3 = 0 \\
\Delta_3 &= \Delta_3' - A\bar{x}R_1 + A\bar{y}R_2 + AR_3 = 0
\end{aligned} \tag{5.54}$$

95

where:

$$A = \int_0^Q \frac{ds}{EI}; \qquad A\bar{x} = \int_0^Q \frac{x\,ds}{EI}; \qquad A\bar{y} = \int_0^Q \frac{y\,ds}{EI}$$

$$I_{xx} = \int_0^Q \frac{y^2 ds}{EI}; \qquad I_{yy} = \int_0^Q \frac{x^2\,ds}{EI}; \qquad I_{xy} = \int_0^Q \frac{xy\,ds}{EI}$$

(5.55)

and Δ'_1, Δ'_2 and Δ'_3 are as in equation (5.45).

Having regard to the form of equations (5.54) it is evident that if the origin is chosen to be at the centre of elastic area ('centre of gravity') as shown in

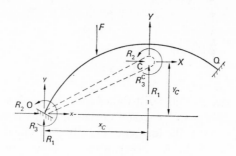

Fig. 5.16

Fig. 5.16, then the terms containing $A\bar{x}$ and $A\bar{y}$ disappear to give only two simultaneous equations and an independent third equation. That is:

$$\Delta_1^c + I_{YY}R_1 - I_{XY}R_2 = 0$$
$$\Delta_2^c - I_{XY}R_1 + I_{XX}R_2 = 0$$
$$\Delta_3^c + AR_3^c = 0$$

(5.56)

where:

$$\Delta_1^c = -\int_0^Q \frac{M_0 X\,ds}{EI}; \qquad \Delta_2^c = \int_0^Q \frac{M_0 Y\,ds}{EI}; \qquad \Delta_3^c = \int_0^Q \frac{M_0\,ds}{EI} = \Delta'_3$$

$$I_{XX} = \int_0^Q \frac{Y^2\,ds}{EI}; \qquad I_{YY} = \int_0^Q \frac{X^2\,ds}{EI}; \qquad I_{XY} = \int_0^Q \frac{XY\,ds}{EI}$$

(5.57)

$$R_3^c = R_3 - R_1 x_c + R_2 y_c$$

X, Y being the coordinates of a point on the arch with reference to the new origin and $x_c = \bar{x}$; $y_c = \bar{y}$. Furthermore, if the principal axes are known, as they are for a symmetrical arch and the coordinates (X, Y) are with respect to these axes then $I_{XY} = 0$ and three independent equations each in one unknown are obtained as follows:

$$\Delta_1^c + I_{YY}R_1 = 0$$
$$\Delta_2^c + I_{XX}R_2 = 0$$
$$\Delta_3 + AR_3^c = 0$$

(5.58)

These equations represent a 'normal' set and their physical implication is that forces applied to the arch through the elastic centre with their lines of action coincident with the principal axes there, cause deflections in their own lines of action, respectively. The equations also lead directly to the so-called 'column analogy' due to Hardy Cross[1].

In the particular instances of arches pin-jointed to abutments the pin-joints represent points of infinite elastic area. Thus, if there is a pin at only one abutment the elastic centre coincides with the pin. If, however, there are pins at both abutments, the elastic centre lies midway between the pins on a line through both pins. These are instances, though, for which the concept of the elastic centre is of little value owing to there being in any event, two and one redundants respectively.

5.7 Examples of the use of the concept of elastic centre

(*a*) The simple uniform portal framework with rigid joints loaded as shown in Fig. 5.17 has a flexural rigidity *EI* and its feet are encastré at A and G. Its elastic centre is therefore located on the vertical centre line at, say, a

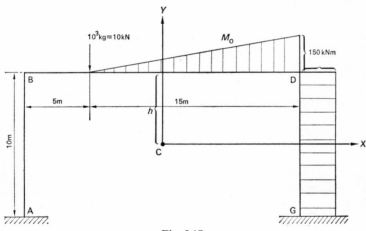

Fig. 5.17

distance *h* m below BD, determined by taking moments of elastic area about BD, as follows:

$$2 \times \frac{10}{EI} \times 5 = \frac{40}{EI} h$$

[1] Equations (5.58) enable the bending moment at any part to be expressed as follows:

$$M = M_0 - \left[\frac{\Delta_3}{A} + \frac{\Delta_1^C}{I_{YY}} X + \frac{\Delta_2^C}{I_{XX}} Y \right]$$

where the term in brackets is analogous to the expression for the stress in any fibre of an eccentrically loaded short column (see bibliography, ref. 6).

4*

whence:

$$h = 2\cdot5 \text{ m}$$

Having thus located C, the line of the axis of X is defined and the flexibility coefficients I_{XX} and I_{XY} are calculated in the following manner, I_{XY} being zero because the axes of X and Y are principal axes:

$$I_{XX} = 2\left[\left(\frac{10^3}{12} + 10 \times 2\cdot5^2\right) + 20 \times 2\cdot5^2\right]/EI = 461\cdot7/EI$$

$$I_{YY} = \left[\frac{20^3}{12} + 2 \times 10^3\right]/EI = 2666\cdot7/EI$$

also, $A = 40/EI$. Then, having regard to the M_0 diagram shown in Fig. 5.17

$$\Delta_1^c = 10\left[\frac{15^2}{2} \times 5 + 150 \times 10\right]/EI = 20\cdot625 \times 10^3/EI$$

$$\Delta_2^c = 10\left[\frac{15^2}{2} \times 2\cdot5 - 150 \times 2\cdot5\right]/EI = -0\cdot938 \times 10^3/EI$$

$$\Delta_3^c = 10\left[\frac{15^2}{2} + 15 \times 10\right]/EI = 2\cdot63 \times 10^3/EI$$

Therefore:

$$R_1 = \frac{\Delta_1^c}{I_{YY}} = \frac{20\cdot625 \times 10^3}{2666\cdot7} = 7\cdot74 \text{ kN}$$

$$R_2 = -\frac{\Delta_2^c}{I_{XX}} = \frac{0\cdot938 \times 10^3}{416\cdot7} = 2\cdot25 \text{ kN}$$

$$R_3^c = -\frac{\Delta_3^c}{A} = -\frac{2\cdot63 \times 10^3}{40} = -65\cdot75 \text{ kN m}$$

from which the bending moments at A, B, D and G are obtained as follows, by taking moments about these points, respectively:

$$M_A = -(2\cdot25 \times 7\cdot5) + (7\cdot74 \times 10) - 65\cdot75 = -5\cdot29 \text{ kN m}$$

$$M_B = (2\cdot25 \times 2\cdot5) + (7\cdot74 \times 10) - 65\cdot75 = 17\cdot28 \text{ kN m}$$

$$M_D = 150 + (2\cdot25 \times 2\cdot5) - (7\cdot74 \times 10) - 65\cdot75 = 12\cdot48 \text{ kN m}$$

$$M_G = 150 - (2\cdot25 \times 7\cdot5) - (7\cdot74 \times 10) - 65\cdot75 = -10.03 \text{ kN m}$$

The positive sign represents a bending moment tending to hogg or close up the framework. The complete bending moment diagram for the framework is shown in Fig. 5.18.

It is instructive to solve this problem alternatively using the symmetrical and anti-symmetrical components of the loading. The latter causes equal uniform bending moments in the stanchions because the horizontal reaction at their feet is zero (para. 5.4). If then the stanchion feet are pin-jointed to rigid foundations there is no bending of the stanchions for this component of loading.

(*b*) The presence of flexibility in bending at foundations merely involves taking the corresponding additional elastic area into account in order to employ the elastic centre method. Thus, if the framework is connected to

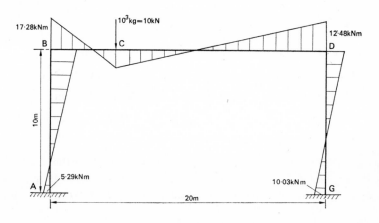

Fig. 5.18

rigid foundations by joints whose flexibility in bending is $10/EI$ this additional amount of elastic area must be taken into account as being concentrated at both A and G. For the position of the elastic centre C, therefore, moments of elastic area about BD gives:

$$2 \times 10 \times 5 + 2 \times 10 \times 10 = (40 + 20)h$$

whence:

$$h = 5 \cdot 0 \text{ m}$$

Also, now:

$$A = (40 + 20)/EI = 60 \cdot 0/EI$$

$$I_{XX} = (2 \times 10^3/12 + 20 \times 5^2 + 2 \times 10 \times 5^2)/EI = 1166 \cdot 7/EI$$

$$I_{YY} = (20^3/12 + 10^3 \times 2 + 10^3 \times 2)/EI = 4666 \cdot 7/EI$$

$$\Delta_1^C = (5 \times 15^2/2 + 150 \times 10 + 15 \times 10 \times 10)10/EI$$
$$= 35 \cdot 626 \times 10^3/EI$$

$$\Delta_2^C = (5 \times 15^2/2 - 15 \times 5 \times 10)10/EI = -1 \cdot 875 \times 10^3/EI$$

$$\Delta_3^C = (15^2/2 + 15 \times 10 + 15 \times 10)10/EI = 4 \cdot 125 \times 10^3/EI$$

99

Therefore:

$$R_1 = \frac{\Delta_1^C}{I_{YY}} = \frac{35626}{4666 \cdot 7} = 7 \cdot 63 \text{ kN}$$

$$R_2 = -\frac{\Delta_2^C}{I_{XX}} = \frac{1875}{1166 \cdot 7} = 1 \cdot 61 \text{ kN}$$

$$R_3^C = -\frac{\Delta_3^C}{A} = -\frac{4125}{60} = -68 \cdot 75 \text{ kN m}$$

and the bending moments at A, B, D and G are as follows:

$$M_A = -(1 \cdot 61 \times 5) + (7 \cdot 63 \times 10) - 68 \cdot 75 = -0 \cdot 49 \text{ kN m}$$
$$M_B = (1 \cdot 61 \times 5) + (7 \cdot 63 \times 10) - 68 \cdot 75 = 15 \cdot 59 \text{ kN m}$$
$$M_D = 150 + (1 \cdot 61 \times 5) - (7 \cdot 63 \times 10) - 68 \cdot 75 = 12 \cdot 99 \text{ kN m}$$
$$M_G = 150 - (1 \cdot 61 \times 5) - (7 \cdot 63 \times 10) - 68 \cdot 75 = -11 \cdot 12 \text{ kN m}$$

(*c*) In order to make use of the elastic centre method for the analysis of a closed ring system, as shown in Fig. 5.19 it is merely necessary to imagine any section cut as shown there and proceed in accordance with the principles

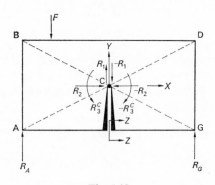

Fig. 5.19

described. For a ring which is symmetrical in every respect the location of the elastic centre coincides with the geometrical centre so that use of the elastic centre method is particularly convenient, especially as the principal axes of the system can also be determined by inspection.

5.8 Choice of statically determinate bending moment

Alternative choices of the distribution of the statically determinate bending moment M_0 are available. Thus, for example, referring to the problem of para. 5.7(*a*) the statically determinate bending moment distribution shown in

Fig. 5.20 may be used. The explanation of this transformation is shown in Fig. 5.21. The introduction of an upward force of 7·5 kN at A of the statically determinate system is permissible if the values of the statically indeterminate

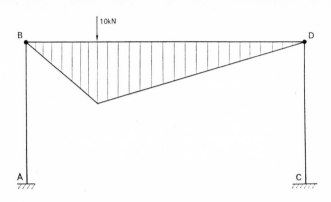

Fig. 5.20

quantities R_1 and R_3^C are modified accordingly. By using the distribution of M_0 shown in Fig. 5.20, such modification is implied as the following solution, corresponding to that of para. 5.7(a) indicates:

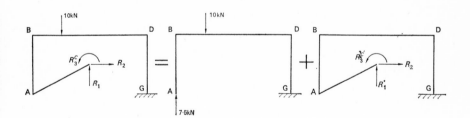

Fig. 5.21

$$\Delta_1^{C'} = 10\left[\frac{5}{2} \times \frac{15}{4}\left(5 + \frac{5}{3}\right)\right]/EI = 625/EI$$

$$\Delta_2^{C'} = -10\left(\frac{20}{2} \times \frac{15}{4} \times \frac{5}{2}\right)/EI = -938/EI$$

$$\Delta_3^{C'} = -10\left(\frac{20}{2} \times \frac{15}{4}\right)/EI = -380/EI$$

Principles of structural analysis

whence:

$$R_1' = \frac{\Delta_1^{C'}}{I_{YY}} = \frac{625}{2666 \cdot 7} = 0 \cdot 24 \text{ kN}$$

$$R_2' = -\frac{\Delta_2^{C'}}{I_{XX}} = \frac{938}{416 \cdot 7} = 2 \cdot 25 \text{ kN}$$

$$R_3^{C'} = -\frac{\Delta_3^{C'}}{A} = \frac{380}{40} = 9 \cdot 38 \text{ kN m}$$

That is, as would be expected $R_1' = (R_1 - 7 \cdot 5)$ kN and $R_3^{C'} = (R_3^C + 75)$ kN m. Also:

$$M_A = -(2 \cdot 25 \times 7 \cdot 5) + (0 \cdot 24 \times 10) + 9 \cdot 38 = -5 \cdot 29 \text{ kN m}$$

and similarly for M_B, M_D and M_G. The kind of transformation illustrated in Fig. 5.21 is of general utility for the application of the compatibility approach in framework analysis of this kind. It is sometimes referred to as the principle of 'mixed systems'.

5.9 Another use of the elastic centre: the flexibility of plane piping systems

A plane pipe-run with rigid anchorages is representative of an elastic arch as shown in Fig. 5.22. If such a system is erected cold and subsequently used for the transmission of hot liquid or gas such as steam, its tendency to expand

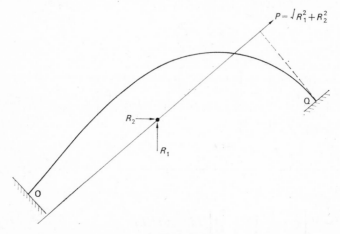

Fig. 5.22

will give rise to forces at the anchorages resisting expansion and consequently, additional stresses will be set up in the pipe wall. Thus, if the end O is perfectly free it will expand by $\Delta_1' = \beta(y_0 - y_Q)\theta$ in the y direction and $\Delta_2' = \beta(x_0 - x_Q)\theta$ in the x direction due to a uniform rise in temperature of θ, where β is the coefficient of linear expansion of the pipe. The forces R_1, R_2 and R_3 at O exerted by the anchorage to prevent expansion are, therefore, such that:

$$R_1 I_{yy} - R_2 I_{xy} - R_3 A \bar{x} + \Delta_1' = 0$$
$$-R_1 I_{xy} + R_2 I_{xx} + R_3 A \bar{y} + \Delta_2' = 0 \qquad (5.59)$$
$$-R_1 A \bar{x} + R_2 A \bar{y} + R_3 A = 0$$

and by changing to the elastic centre as origin:

$$R_1 I_{YY} - R_2 I_{XY} + \Delta_1' = 0$$
$$-R_1 I_{XY} + R_2 I_{XX} + \Delta_2' = 0 \qquad (5.60)$$
$$R_3^C = 0$$

It is interesting that now R_3^C is zero which means that the thrust line for the pipe passes through the elastic centre C, as shown in Fig. 5.22.

In the event of the pipe having branches, it is no longer possible to derive advantage from the concept of the elastic centre nor is it possible to do so if the pipe-run is three dimensional unless it happens that the flexural and torsional rigidities of the pipe are equal. Spielvogel[1] has, however, proposed an approximate method of dealing with three dimensional runs without branches, using a fictitious elastic centre. The implication of this method is that a three-dimensional single run has a simple thrust line whereas, unless the flexural and torsional rigidities are equal, the thrust line has, in fact, a couple acting about it.

This kind of problem is of considerable interest to power plant designers and engineers. Moreover, it illustrates a method of dealing with temperature and other self-straining effects in frameworks of the single-ring type.

5.10 Calculation of deflections of statically indeterminate beam systems and rigidly jointed frameworks

The deflection at any point P of a span of a statically indeterminate beam system, as shown in Fig. 5.23, due to any conditions of bending moment described by M anywhere in the span, may be found readily by means of the principle of virtual work in the manner described in para. 4.5 as follows:

$$1 \Delta_P = \int_0^l M' \frac{M \, dx}{EI} \qquad (5.61)$$

where 1 and M' represent the system of forces in equilibrium shown in Fig. 5.23(b) and $\delta \Delta_P$ and $M \, \delta x / EI$ are compatible virtual displacements equal to those caused by the loading.

[1] See bibliography, ref. 7.

Similarly, the deflection of any point of a simple portal frame such as that loaded as shown in Fig. 5.24(a) in which the bending moment at any point is described by M, can be found by a virtual work equation such as (5.61). Thus, for the vertical deflection due to bending, at any point P of BC, the

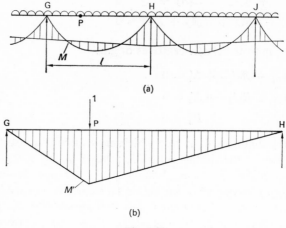

Fig. 5.23

system of forces in equilibrium shown in Fig. 5.24(b) is appropriate so that equation (5.61) may be used, in which $l = BC$. For the horizontal deflection of the tops of the stanchions B and C, the appropriate system of forces in equilibrium is shown in Fig. 5.24(c) and its virtual work due to a system of compatible virtual displacements equal to those of the framework produced by the actual loading, is:

$$1\Delta_B = \int_0^h M'' \frac{M \, ds}{EI} \tag{5.62}$$

As a numerical example, suppose having analysed the portal shown in Fig. 5.17, with flexible foundations by the elastic centre method as described in para. 5.7 the horizontal deflection of the joint B is required. The system in equilibrium shown in Fig. 5.24(c) and the virtual work equation (5.62) are appropriate and for the purpose of the latter:

$$M'' = (10 - y) \text{ kN m}$$

$$M = M_A + \frac{M_B - M_A}{10} y = 0.490 + 1.610y$$

(see para. 5.7). Therefore:

$$1\Delta_B = \left[-0.490 \int_0^{10} (10 - y) \, dy + 1.610 \int_0^{10} (10 - y)y \, dy \right] / EI$$

$$= 240/EI$$

104

where the positive sign indicates deflection in the sense of the fictitious unit load at B as shown in Fig. 5.24(c).

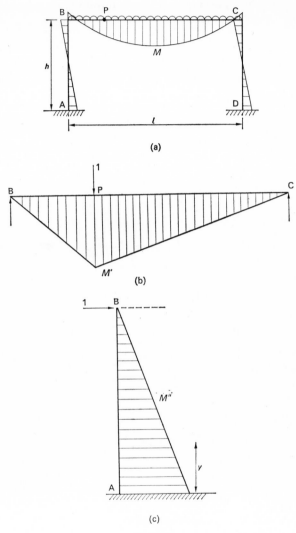

Fig. 5.24

5.11 Conditions of equilibrium within beams and portal frameworks

The conditions of equilibrium within beams and rigidly jointed frameworks may be determined conveniently by applying the principle of virtual work. If, for example, it is required to find the relationship between the load and the bending moments at A, B and C for the encastré beam loaded as shown in Fig. 5.25. B is imagined to suffer a small virtual displacement Δ_B such that the portion AB of the beam suffers virtual rotation through small angle ψ_{AB} while the portion BC suffers virtual rotation through small angle ψ_{BC}. The resulting virtual work is, then:

$$M_A\psi_{AB} + M_{BA}\psi_{AB} + M_{BC}\psi_{BC} + M_C\psi_{BC} + F_B\Delta_B = 0 \tag{5.63}$$

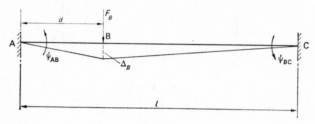

Fig. 5.25

Substituting the conditions:

$$M_{BA} + M_{BC} = 0 \quad \text{or} \quad M_{BA} = -M_{BC} = M_B \text{ (for equilibrium at B)}$$

$$\psi_{BC}(l - d) = -\psi_{AB}d = \Delta_B \text{ (for compatibility)} \tag{5.64}$$

gives finally:

$$F_B = \frac{1}{d}M_A + \frac{l}{d(l-d)}M_B - \frac{1}{l-d}M_C \tag{5.65}$$

Again, for the simple portal shown in Fig. 5.26 a relationship between the bending moments M_A, M_B, M_C and M_D at the extremities of the stanchions

Fig. 5.26

(i.e. couples applied to the ends of the stanchions) and the loading can be found by considering virtual sway of the framework through a small angle ψ as shown, when by virtual work:

$$(M_A + B_B + M_C + M_D)\psi - F_B\psi h = 0 \tag{5.66}$$

or

$$M_A + M_B + M_C + M_D = F_B h \tag{5.67}$$

106

In the event of, say, the stanchion AB being inclined at 60° to the horizontal as shown in Fig. 5.27(a) the virtual work equation for a virtual sway of the framework now contains the bending moments $M_{BC} = -M_B$ and $M_{CB} = -M_C$ at the ends of the beam as well. This is due to the virtual rotation of the

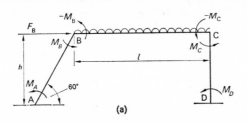

(a)

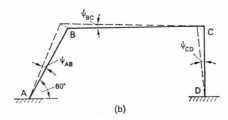

(b)

Fig. 5.27

beam which occurs during the sway and for compatibility of the virtual rotations of the members (Fig. 5.27(b)), clearly:

$$h\psi_{AB} = h\psi_{CD} \text{ (horizontal displacement of B and C)}$$

or   (5.68)

$$\psi_{AB} = \psi_{CD} = \psi$$

and:

$$\frac{h}{\sqrt{3}}\psi_{AB} = \frac{h}{\sqrt{3}}\psi = -h\psi_{BC} \text{ (vertical displacement of B)}$$

or

$$\psi_{BC} = -\frac{h}{l\sqrt{3}} \tag{5.69}$$

Application of the principle of virtual work to the sway now gives:

$$M_A\psi + M_B\psi + \left(M_B \times \frac{h}{l\sqrt{3}}\right) + \left(M_C \times \frac{h}{l\sqrt{3}}\psi\right) + M_C\psi$$

$$+ M_D\psi - F_B\psi h - wl\frac{l}{2}\psi = 0 \tag{5.70}$$

where w is the intensity of the uniformly distributed loading on BC.

107

Therefore, finally

$$M_A + M_B\left(1 + \frac{h}{l\sqrt{3}}\right) + M_C\left(1 + \frac{h}{l\sqrt{3}}\right) + M_D = F_B h + \frac{wl^2}{2} \quad (5.71)$$

This use of virtual work is valuable as a time-saver when it is necessary to determine internal forces and bending moments from loads and calculated values elsewhere in the framework.

EXERCISES

1 A uniform beam ABC of length $2l$ has simple rigid supports at A, B and C. Determine the reactions at the supports due to a uniformly distributed load of intensity w per unit length which extends for a distance of $3l/2$ from A. What would be the reaction at the centre support at B if it was elastic with a flexibility of $l^3/48EI$?

Ans: $R_B = 1\cdot07wl$; $0\cdot95wl$

2 A uniform beam ABC of length $2l$ has simple rigid supports at A and B and is encastré at C. Determine the reactions at the supports due to a uniformly distributed load of intensity w per unit length over AB and a concentrated load of $0\cdot5wl$ acting midway between B and C. AB = BC = l.

Ans: $0\cdot4wl$; $0\cdot9wl$; $0\cdot2wl$; $0\cdot05wl^2$

3 A uniform rigidly jointed plane portal framework ABCD is such that AB = BC = CD = l and the feet A and D of the stanchions are encastré in rigid foundations. Show that when a horizontal load H is applied at B (or C) the bending moment at the foot of each stanchion is $2Hl/7$.

4 A uniform symmetrical pitched roof portal framework ABCDG has rigid joints and stanchion feet encastré in rigid foundations at C and G. The span of the framework is $2l$; the height to eaves is l and the rise of the rafters is $l/2$. Show that for a uniformly distributed load of intensity w per unit horizontal length over the span $2l$ of the rafters, the bending moment at B and D is of magnitude $0\cdot21wl^2$. Show also that the horizontal deflection of B and D is of magnitude $wl^4/432EI$.

5 A uniform arch encastré in rigid abutments is of parabolic shape of span 100 m and rise 25 m. Calculate the horizontal thrust at the abutments due to a uniformly distributed load of 10^5 kg per horizontal metre over the span. What is the distribution of bending moment in the arch?

If the uniformly distributed load extends over only one-half of the span what is then the horizontal thrust at the abutments due to the loading?

Ans: 5×10^4 kN; 0; $2\cdot5 \times 10^4$ kN

6 Verify the information given in Table 6.1 by using the elastic centre concept.

6

Analysis by the equilibrium approach of statically indeterminate beam systems and rigidly jointed frameworks

6.1 The analysis of statically indeterminate beam systems and rigidly jointed frameworks whose elasticity is linear are treated in this chapter, using the equilibrium approach (para. 3.2). This approach is preferable in general for such systems and its application is facilitated by the use of the concept of stiffness coefficients. Having identified the degrees of freedom of a structure the analysis then follows a common pattern which is particularly convenient for automatic computation. In this chapter, as in Chapter 5, it is assumed that axial and shear deformation of structural members is negligible, with the exception of the contents of para. 6.7 concerning secondary effects.

6.2 Some basic relationships for beams

The relationships shown in Fig. 6.1 for a uniform cantilever AB of flexural rigidity EI, encastré at A are essential for deriving the stiffness coefficients of statically indeterminate beam systems and rigidly jointed frameworks made of uniform members. They may be derived by using the data given in Table 4.1. In Fig. 6.1(a) the slope at B caused by unit couple applied there when deflection of B is prevented, is given, together with the magnitude of the restraining force to prevent translational deflection of B. Fig. 6.1(b) shows the system of forces and couples to cause unit slope at B in the absence of translational deflection of that point. Finally, Fig. 6.1(c) shows the system of forces and couples causing unit translational deflection of B in the absence of slope there.

Another relevant set of relationships concerns a simply supported uniform beam whose supports differ in level by a small amount, subjected to terminal

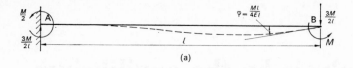

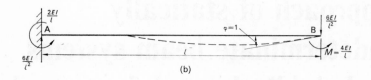

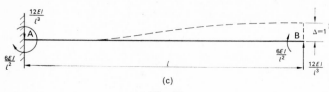

Fig. 6.1

couples as shown in Fig. 6.2. If the resulting end slopes are measured from a horizontal datum, then on the basis of equation (4.26) and the principle of superposition:

$$M_A = \frac{4EI}{l}(\phi_A - \psi) + \frac{2EI}{l}(\phi_B - \psi)$$

$$M_B = \frac{2EI}{l}(\phi_A - \psi) + \frac{4EI}{l}(\phi_B - \psi)$$

(6.1)

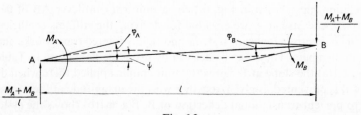

Fig. 6.2

where ψ is the small angle of slope of the beam due to the difference in level of the supports. Equations (6.1), so-called 'slope-deflection equations', are necessary for calculating the terminal bending couples acting on structural

110

members following analysis which provides the terminal slopes and difference in level.

6.3 Treatment of loads on a member

If a beam AB as part of a structure is loaded as shown in Fig. 6.3(a) then it will have no effect on the structure as a whole if external forces and couples of

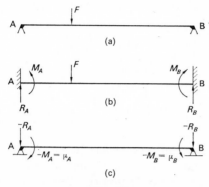

Fig. 6.3

appropriate magnitude are applied at the joints A and B as shown in Fig. 6.3(b). It follows, then, that as far as the structure as a whole is concerned, the effect of the loading within the beam AB is the same as if the system of 'restraining' forces and couples shown in Fig. 6.3(b) is reversed and applied at A and B as shown in Fig. 6.3(c). In other words, superposition of the external force systems of Figs. 6.3(b) and (c) gives the external force system of Fig. 6.3(a).

The magnitude of the restraining forces and couples is clearly identical to the supporting forces and couples which would be necessary to make the loaded beam encastré. The couples are, therefore, the 'fixed-end' moments for the beam, while the restraining forces are such that together they are equal in magnitude but opposite in sense to the beam loading, their individual values

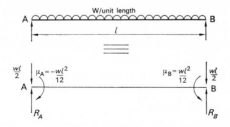

Fig. 6.4

being determined by the requirement of zero resultant couple on the beam as a whole. Values of fixed-end moments and reactions for three components of load on a uniform beam with linear elasticity are shown in Table 6.1, and the 'end slope producing effect' of a uniformly distributed load is shown in Fig. 6.4.

TABLE 6.1

System of loading on a uniform encastré beam AB $= l$	Fixed-end moment at A	Fixed-end moment at B	Vertical reaction at A	Vertical reaction at B
Central concentrated load F	$Fl/8$ anticlockwise	$Fl/8$ clockwise	$F/2$	$F/2$
Uniformly distributed load w per unit length	$wl^2/12$ anticlockwise	$wl^2/12$ clockwise	$wl/2$	$wl/2$
Concentrated load F acting at a distance p from A and q from B	Fpq^2/l^2 anticlockwise	Fp^2q/l^2 clockwise	$Fq^2(q + 3p)/l^3$	$Fp^2(p + 3q)/l^3$

The end slopes of a uniform beam AB due to span loading and external terminal couples M_A and M_B with supporting forces to provide equilibrium, can be obtained by the slope-deflection equations (6.1). However, M_A and M_B must be replaced by $(M_A + \mu_A)$ and $(M_B + \mu_B)$, respectively, where μ_A and μ_B denote the reversed fixed-end moments to take account of the loading within the span of the beam. Thus:

$$M_A + \mu_A = b_{AA}(\phi_A - \psi) + b_{AB}(\phi_B - \psi)$$
$$M_B + \mu_B = b_{BA}(\phi_A - \psi) + b_{BB}(\phi_B - \psi)$$

(6.2)

where the stiffness coefficients b have the values given in equations (6.1).

6.4 Analysis of statically indeterminate beam systems

In common with elastic structures of all kinds, the analysis of beam systems and rigidly jointed frameworks by the equilibrium approach consists of setting-up and solving the relevant equations of equilibrium. For this purpose it is convenient to choose the components of deflection of joints (corresponding to the degrees of freedom of those joints elastically) as unknowns and, for linearly elastic structures, to use the constant stiffness coefficients relevant to the degrees of freedom of the joints to relate the deflections to the applied loading referred to the joints (para. 6.3). This procedure is described in para. 3.2 for trusses and the final form of the equations of equilibrium given there is precisely the same as for beam systems and rigidly jointed frameworks when everything is referred to joints. Its application to beam systems is conveniently illustrated by a few specific examples.

(a) A uniform cantilever AB of flexural rigidity EI and length l has a linearly elastic prop at B whose stiffness is b_1 as shown in Fig. 6.5(a). It is required to analyse the system for a uniformly distributed load of intensity w per unit length over AB. The system has two degrees of elastic freedom of joints namely the vertical and rotational deflection of B represented by Δ_1 and Δ_2, respectively. The loading referred to the joints is such that there is a (downward) force $F_1 = -wl/2$ and a couple $F_2 = wl^2/12$ acting at B.

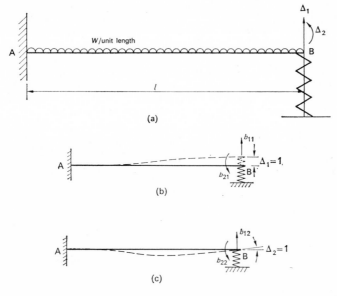

Fig. 6.5

The equations of equilibrium with Δ_1 and Δ_2 as unknowns are, then, as follows:

$$b_{11}\Delta_1 + b_{12}\Delta_2 = F_1$$
$$b_{21}\Delta_1 + b_{22}\Delta_2 = F_2$$
(6.3)

where the stiffness coefficients b_{11}, $b_{12} = b_{21}$ and b_{22} are described in Fig. 6.5(b) and (c) and on the basis of the data of Fig. 6.1 have the values:

$$b_{11} = 12EI/l^3 + b_1; \qquad b_{12} = b_{21} = -6EI/l^2; \qquad b_{22} = 4EI/l \quad (6.4)$$

Therefore, finally:

$$\left(\frac{12EI}{l^3} + b_1\right)\Delta_1 - \frac{6EI}{l^2}\Delta_2 = -\frac{wl}{2}$$
$$-\frac{6EI}{l^2}\Delta_1 + \frac{4EI}{l}\Delta_2 = \frac{wl^2}{12}$$
(6.5)

113

whence:

$$\Delta_1 = -\frac{3wl^4}{8(3EI + b_1l^3)}$$

$$\Delta_2 = \frac{3wl^3}{16}\left(\frac{1}{9EI} - \frac{3}{3EI + b_1l^3}\right) \tag{6.6}$$

(*b*) The uniform continuous beam shown in Fig. 6.6 is supported rigidly at A, B, C and D. There are two redundants, for example, the connections at

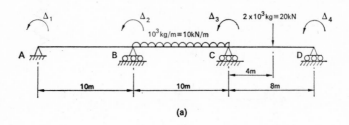

(a)

(b)

Fig. 6.6

B and C or the supports at these points and four degrees of freedom of joints represented by the slope of the beam at the supports. For the loading shown:

$$\mu_{AB} = \mu_{BA} = 0$$

$$\mu_{BC} = -83\cdot3 \text{ kN m}$$

$$\mu_{CB} = 83\cdot3 \text{ kN m} \tag{6.7}$$

$$\mu_{CD} = -20\cdot0 \text{ kN m}$$

$$\mu_{DC} = 20\cdot0 \text{ kN m}$$

and the loads on the supports are 50, 60 and 10 kN at B, C and D, respectively.

The relevant stiffness coefficients in this instance represent the system of couples which cause unit rotational displacement at A, B, C and D, when there is no displacement at the three other points. These concepts are illustrated in Fig. 6.7(a), (b), (c) and (d), respectively. Thus, with reference to Fig. 6.1:

$$b_{11} = \frac{4EI}{10}; \quad b_{21} = \frac{2EI}{10}; \quad b_{31} = 0; \quad b_{41} = 0$$

$$b_{12} = \frac{2EI}{10}; \quad b_{22} = 4EI\left(\frac{1}{10} + \frac{1}{10}\right); \quad b_{32} = \frac{2EI}{10}; \quad b_{42} = 0$$

$$b_{13} = 0; \quad b_{23} = \frac{2EI}{10}; \quad b_{33} = 4EI\left(\frac{1}{10} + \frac{1}{8}\right); \quad b_{43} = \frac{2EI}{8}$$

$$b_{14} = 0; \quad b_{24} = 0; \quad b_{34} = \frac{2EI}{8}; \quad b_{44} = \frac{4EI}{8}$$

(6.8)

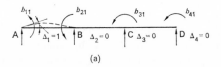

(a)

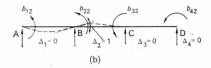

(b)

(c)

(d)

Fig. 6.7

and for equilibrium:

$$b_{11}\Delta_1 + b_{12}\Delta_2 + b_{13}\Delta_3 + b_{14}\Delta_4 = F_1$$
$$b_{21}\Delta_1 + b_{22}\Delta_2 + b_{23}\Delta_3 + b_{24}\Delta_4 = F_2$$
$$b_{31}\Delta_1 + b_{32}\Delta_2 + b_{33}\Delta_3 + b_{34}\Delta_4 = F_3$$
$$b_{41}\Delta_1 + b_{42}\Delta_2 + b_{43}\Delta_3 + b_{44}\Delta_4 = F_4$$

(6.9)

where $F_1 = \mu_{AB} = 0$; $F_2 = \mu_{BA} + \mu_{BC} = -83.3$ kN m; $F_3 = \mu_{CB} + \mu_{CD} = 63.3$ kN m; $F_4 = \mu_{DC} = 20.0$ kN m, by equation (6.7). Substituting these

115

values and the values of the stiffness coefficients (6.8) in equations (6.9) gives

$$
\begin{aligned}
EI(0\cdot400\Delta_1 + 0\cdot200\Delta_2 + 0\Delta_3 \quad + 0\Delta_4) \quad &= 0 \\
EI(0\cdot200\Delta_1 + 0\cdot800\Delta_2 + 0\cdot200\Delta_3 + 0\Delta_4) \quad &= -83\cdot3 \\
EI(0\Delta_1 \quad + 0\cdot200\Delta_2 + 0\cdot900\Delta_3 + 0\cdot250\Delta_4) &= 63\cdot3 \\
EI(0\Delta_1 \quad + 0\Delta_2 \quad + 0\cdot250\Delta_3 + 0\cdot500\Delta_4) &= 20\cdot0
\end{aligned}
\tag{6.10}
$$

whence, in radians:

$$
\begin{aligned}
\Delta_1 = \phi_A = 79\cdot8/EI; \qquad &\Delta_2 = \phi_B = -159\cdot6/EI; \\
\Delta_3 = \phi_C = 106\cdot9/EI; \qquad &\Delta_4 = \phi_D = 13\cdot6/EI
\end{aligned}
\tag{6.11}
$$

The bending moments at B and C can be found from these deflections by means of equations (6.2).

As the span AB in this example is unloaded it is possible to eliminate one degree of freedom, namely the one at A, by calculating the stiffness coefficients relating to B on the basis of a simple support at A. Thus $b_{11}, b_{12} = b_{21}$ and Δ_1 no longer appear and b_{22} is modified to $EI(\frac{3}{10} + \frac{4}{10})$ (see Fig. 6.8(a))

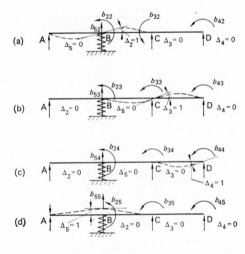

Fig. 6.8

to take account of the rotational freedom at A. Finally, the three equations of equilibrium are identical to the second, third and fourth of equations (6.10) except for the modified value of b_{22} and the elimination of the Δ_1 terms.

(c) If, say, the support at B of the continuous beam, considered in (b) above, sinks elastically under load an additional degree of freedom is introduced. In order to illustrate the treatment to include this feature suppose $EI = 2 \times 10^4$ kNm^2 and the stiffness of the support at B, $b_5 = 10^3$ kN/m.

The stiffness coefficients are calculated on the basis of the conditions shown in Fig. 6.8(a), (b), (c) and (d) as follows:

$$b_{22} = EI\left(\frac{3}{10} + \frac{4}{10}\right); \quad b_{32} = \frac{2EI}{10}; \quad b_{42} = 0; \quad b_{52} = EI\left(\frac{6}{10^3} - \frac{3}{10^3}\right)$$

$$b_{23} = \frac{2EI}{10}; \quad b_{33} = 4EI\left(\frac{1}{10} + \frac{1}{8}\right); \quad b_{43} = \frac{2EI}{8}; \quad b_{53} = \frac{6EI}{10^2}$$

$$b_{24} = 0; \quad b_{34} = \frac{2EI}{8}; \quad b_{44} = \frac{4EI}{8}; \quad b_{54} = 0 \qquad (6.12)$$

$$b_{25} = EI\left(\frac{6}{10^2} - \frac{3}{10^2}\right); \quad b_{35} = \frac{6EI}{10^2};$$

$$b_{45} = 0; \quad b_{55} = b_5 + \frac{3EI}{10^3} + \frac{12EI}{10^3}$$

Substituting the values for EI and b_5 given above and remembering that $F_5 = -50$ kN due to the loading on BC, the final equations of equilibrium are as follows:

$$
\begin{aligned}
1400\Delta_2 + 400\Delta_3 + 0\Delta_4 + 60\Delta_5 &= -8{\cdot}333 \\
400\Delta_2 + 1800\Delta_3 + 500\Delta_4 + 120\Delta_5 &= 6{\cdot}333 \\
0\Delta_2 + 500\Delta_3 + 1000\Delta_4 + 0\Delta_5 &= 2{\cdot}000 \\
60\Delta_2 + 120\Delta_3 + 0\Delta_4 + 130\Delta_5 &= -5{\cdot}000
\end{aligned}
\qquad (6.13)
$$

whence:

$$\Delta_2 = -0{\cdot}0065 \text{ rad}; \quad \Delta_3 = 0{\cdot}0085 \text{ rad};$$
$$\Delta_4 = -0{\cdot}0043 \text{ rad}; \quad \Delta_5 = -0{\cdot}0433 \text{ m}$$
$$(6.14)$$

from which the bending moments at B and C can be found by means of equations (6.2). The solution of equations (6.10) and (6.13) is simplified because there are several zero coefficients; in general, however, solution of simultaneous equations is conveniently accomplished by the method shown in para. 6.6 (see also paras. 9.3 and 9.5).

6.5 Analysis of statically indeterminate, rigidly jointed frameworks

The essential features of the analysis of rigidly jointed frames can be appreciated by considering a simple portal frame of uniform section and with stanchion feet encastré in rigid foundations, as shown in Fig. 6.9.

Assuming that axial effects are negligible, this frame has three degrees of freedom and three redundants. Thus, the joints at B and C are capable of rotation and the frame as a whole can sway sideways while the rigid connections at B and C and at either A and D in respect of rotation, are redundant.

Principles of structural analysis

The final equations of equilibrium for the frame can be written generally as
follows:

$$b_{11}\Delta_1 + b_{12}\Delta_2 + b_{13}\Delta_3 = F_1$$
$$b_{21}\Delta_1 + b_{22}\Delta_2 + b_{23}\Delta_3 = F_2 \qquad (6.15)$$
$$b_{31}\Delta_1 + b_{32}\Delta_2 + b_{33}\Delta_3 = F_3$$

where Δ_1 (the sway deflection) $\Delta_2, \Delta_3, F_1, F_2$ and F_3 are as shown in Fig. 6.9(b).

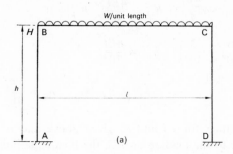

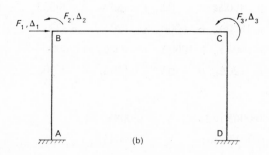

Fig. 6.9

In order to calculate the stiffness coefficients it is necessary to consider the
existence alone and in turn of unit values of Δ_1, Δ_2 and Δ_3 as shown in Fig.
6.10. Thus b_{11} is the force applied horizontally at B which produces unit sway

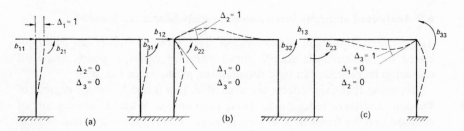

Fig. 6.10

118

displacement when rotation of B and C is prevented by couples b_{21} and b_{31}, that is, by referring to Fig. 6.1

$$b_{11} = \frac{24EI}{h^3}; \qquad b_{21} = \frac{6EI}{h^2} = b_{12}; \qquad b_{31} = \frac{6EI}{h^2} = b_{13} \qquad (6.16)$$

Similarly, b_{22} is the couple which produces unit rotational displacement of B

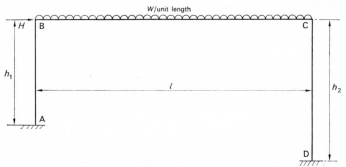

Fig. 6.11

when sway of the frame and rotation of C are prevented by a force of b_{12} and a couple of b_{32}, respectively, that is:

$$b_{22} = 4EI\left(\frac{1}{h} + \frac{1}{l}\right); \qquad b_{32} = \frac{2EI}{l} = b_{23} \qquad (6.17)$$

Finally, b_{33} is the couple which produces unit rotational displacement of C when sway of the frame and rotation of B are prevented by a force of b_{13} and a couple of b_{23}, respectively, and is identical to b_{22} in this instance due to symmetry. Substituting these values of the coefficients and $F_1 = H$, $F_2 = \mu_{BC} = -wl^2/12$ and $F_3 = \mu_{CB} = wl^2/12$ in equations (6.15) gives:

$$\frac{24EI}{h^3}\Delta_1 + \frac{6EI}{h^2}\Delta_2 + \frac{6EI}{h^2}\Delta_3 = H$$

$$\frac{6EI}{h^2}\Delta_1 + 4EI\left(\frac{1}{h} + \frac{1}{l}\right)\Delta_2 + \frac{2EI}{l}\Delta_3 = -\frac{wl^2}{12} \qquad (6.18)$$

$$\frac{6EI}{h^2}\Delta_1 + \frac{2EI}{l}\Delta_2 + 4EI\left(\frac{1}{h} + \frac{1}{l}\right)\Delta_3 = \frac{wl^2}{12}$$

whereby Δ_1, Δ_2 and Δ_3 can be found.

In the event of the stanchions being different in length as shown in Fig. 6.11 the values of the stiffness coefficients are as follows:

$$b_{11} = 12EI\left(\frac{1}{h_1^3} + \frac{1}{h_2^3}\right); \qquad b_{21} = \frac{6EI}{h_1^2} = b_{12}; \qquad b_{31} = \frac{6EI}{h_2^2};$$

$$b_{22} = 4EI\left(\frac{1}{h_1} + \frac{1}{l}\right); \qquad b_{32} = b_{23} = \frac{2EI}{l}; \qquad (6.19)$$

$$b_{33} = 4EI\left(\frac{1}{h_2} + \frac{1}{l}\right)$$

otherwise the analysis is unchanged.

6.6 Examples of the analysis of rigidly jointed frameworks

(*a*) The two-bay, single storey frame, shown in Fig. 6.12, has four degrees of freedom being, the rotational freedom of the joints at B, C and D, and the sidesway freedom of the frame as a whole, and six redundants. For the purpose of using the equilibrium approach the stiffness coefficients are as set out in Table 6.2 below, assuming that *EI* is the same for all members of the frame.

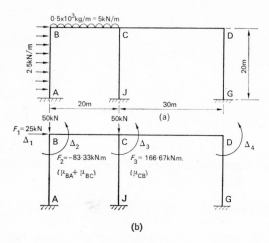

Fig. 6.12

TABLE 6.2

$b \rightarrow$ ↓	1	2	3	4
1	$3 \cdot \dfrac{12EI}{h^3}$ $= 0{\cdot}0045EI$	Since $b_{21} = b_{12}$ $0{\cdot}0150EI$	Since $b_{31} = b_{13}$ $0{\cdot}0150EI$	Since $b_{41} = b_{14}$ $0{\cdot}0150EI$
2	$\dfrac{6EI}{h^2}$ $= 0{\cdot}0150EI$	$4EI\left(\dfrac{1}{h} + \dfrac{1}{l_1}\right)$ $= 0{\cdot}4000EI$	Since $b_{32} = b_{23}$ $0{\cdot}1000EI$	Since $b_{42} = b_{24}$ 0
3	$\dfrac{6EI}{h^2}$ $= 0{\cdot}0150EI$	$\dfrac{2EI}{l_1}$ $= 0{\cdot}1000EI$	$4EI\left(\dfrac{1}{h} + \dfrac{1}{l_1} + \dfrac{1}{l_2}\right)$ $= 0{\cdot}5333EI$	Since $b_{43} = b_{34}$ $0{\cdot}0667EI$
4	$\dfrac{6EI}{h^2}$ $= 0{\cdot}0150EI$	0	$\dfrac{2EI}{l_2}$ $= 0{\cdot}0667EI$	$4EI\left(\dfrac{1}{h} + \dfrac{1}{l_2}\right)$ $= 0{\cdot}3333EI$

$h = $ AB; $l_1 = $ BC; $l_2 = $ CD

120

The equivalent loads for the system of loading shown in Fig. 6.12 are: $F_1 = 25$ kN; $F_2 = -83\cdot33$ kN m; $F_3 = 166\cdot67$ kN m; $F_4 = 0$. Therefore, the final equations of equilibrium are:

$$EI(0\cdot0045\Delta_1 + 0\cdot0150\Delta_2 + 0\cdot0150\Delta_3 + 0\cdot0150\Delta_4) = 25\cdot00$$
$$EI(0\cdot0150\Delta_1 + 0\cdot4000\Delta_2 + 0\cdot1000\Delta_3 + \quad 0\Delta_4) = -83\cdot33$$
$$EI(0\cdot0150\Delta_1 + 0\cdot1000\Delta_2 + 0\cdot5333\Delta_3 + 0\cdot0667\Delta_4) = 166\cdot67$$
$$EI(0\cdot0150\Delta_1 + 0\Delta_2 \quad + 0\cdot0667\Delta_3 + 0\cdot3333\Delta_4) = 0$$

$$(6.20)$$

which can be solved by the Gauss method shown in Table 6.3 as follows:

TABLE 6.3

Operation	Δ_1	Δ_2	Δ_3	Δ_4	$F \times EI$
(i)	0·0045	0·0150	0·0150	0·0150	25·00
(ii)	0·0150	0·4000	0·1000	0	−83·33
(i) × 0·0150/0·0045 = (i)′	0·0150	0·0500	0·0500	0·0500	83·33
(ii)′ = (ii) − (i)′	0	0·3500	0·0500	−0·0500	−166·67
(iii)	0·0150	0·1000	0·5333	0·0667	166·67
(i)′	0·0150	0·0500	0·0500	0·0500	83·33
(iii)′ = (iii) − (i)′	0	0·0500	0·4833	0·0167	83·33
(ii) × 0·05/0·35 = (ii)″		0·0500	0·0071	−0·0071	−23·81
(iii)″ = (iii)′ − (ii)″		0	0·4762	0·0238	107·14
(iv)	0·0150	0	0·0667	0·3333	0
(i)′	0·0150	0·0500	0·0500	0·0500	83·33
(iv)′ = (iv) − (i)′	0	−0·0500	0·0167	0·2833	−83·33
(ii)″		0·0500	0·0071	−0·0071	−23·81
(iv)″ = (iv)′ + (ii)″		0	0·0238	0·2762	−107·14
(iii)″ × 0·0238/0·4762 = (iii)″			0·0238	0·0012	5·35
(iv)″ = (iv)″ − (iii)″			0	0·2750	−112·49

$$\Delta_4 = -4090/EI$$

By substitution back in the process:

$$\Delta_1 = 8000/EI \text{ m}; \quad \Delta_2 = -569/EI \text{ rad}; \quad \Delta_3 = 246/EI \text{ rad.} \quad (6.21)$$

(EI in kN m²).

In the event of one or more of the stanchion feet being pinned to the rigid foundations additional stiffness coefficients in respect of these additional degrees of rotational freedom can be calculated. Alternatively, if the stanchions themselves are not loaded the number of coefficients is unchanged if the pinned connections are taken into account for their determination in the manner described in para. 6.4(c). In the former instance the number of final equations is increased by as many as three while in the latter it is the same as above, namely, four.

TABLE 6.4

$b \rightarrow$	1	2	3	4	5	6
1	$2 \times \dfrac{12EI}{h_2^3}$ $= 0\cdot0030EI$	Since $b_{21} = b_{12}$ $0\cdot0150EI$	Since $b_{31} = b_{13}$ $0\cdot0150EI$	Since $b_{41} = b_{14}$ $-0\cdot0030EI$	Since $b_{51} = b_{15}$ $0\cdot0150EI$	Since $b_{61} = b_{16}$ $0\cdot0150EI$
2	$\dfrac{6EI}{h_2^2}$ $= 0\cdot0150EI$	$4EI\left(\dfrac{1}{l} + \dfrac{1}{h_2}\right)$ $= 0\cdot3333EI$	Since $b_{32} = b_{23}$ $0\cdot0667EI$	Since $b_{42} = b_{24}$ $-0\cdot0150EI$	Since $b_{52} = b_{25}$ $0\cdot1000EI$	Since $b_{62} = b_{26}$ 0
3	$\dfrac{6EI}{h_2^2}$ $= 0\cdot0150EI$	$\dfrac{2EI}{l}$ $= 0\cdot0667EI$	$4EI\left(\dfrac{1}{l} + \dfrac{1}{h_2}\right)$ $= 0\cdot3333EI$	Since $b_{43} = b_{34}$ $-0\cdot0150EI$	Since $b_{53} = b_{35}$ 0	Since $b_{63} = b_{36}$ $0\cdot1000EI$
4	$-2 \times \dfrac{12EI}{h_2^3}$ $= -0\cdot0030EI$	$-\dfrac{6EI}{h_2^2}$ $= -0\cdot0150EI$	$-\dfrac{6EI}{h_2^2}$ $= -0\cdot0150EI$	$2 \times 12EI\left(\dfrac{1}{h_1^3} + \dfrac{1}{h_2^3}\right)$ $= 0\cdot0039EI$	Since $b_{54} = b_{45}$ $-0\cdot0083EI$	Since $b_{64} = b_{46}$ $-0\cdot0083EI$
5	$\dfrac{6EI}{h_2^2}$ $= 0\cdot0150EI$	$\dfrac{2EI}{h_2}$ $= 0\cdot1000EI$	0	$6EI\left(\dfrac{1}{h_1^2} - \dfrac{1}{h_2^2}\right)$ $= -0\cdot0083EI$	$4EI\left(\dfrac{2}{l} + \dfrac{1}{h_1} + \dfrac{1}{h_2}\right)$ $= 0\cdot6000EI$	Since $b_{65} = b_{56}$ $0\cdot1333EI$
6	$\dfrac{6EI}{h_2^2}$ $= 0\cdot0150EI$	0	$\dfrac{2EI}{h_2}$ $= 0\cdot1000EI$	$6EI\left(\dfrac{1}{h_1^2} - \dfrac{1}{h_2^2}\right)$ $= -0\cdot0083EI$	$2EI\left(\dfrac{2}{l}\right)$ $= 0\cdot1333EI$	$4EI\left(\dfrac{2}{l} + \dfrac{1}{h_1} + \dfrac{1}{h_2}\right)$ $= 0\cdot6000EI$

Note: the values in this table are for lengths in dm.
h_1 = AB; h_2 = BC; l = BG = CD

(*b*) The framework shown in Fig. 6.13 has six degrees of freedom; joints B, C, D and G can rotate and each storey can sway sideways independently. The number of redundants is also six. The stiffness coefficients are set out in Table 6.4 on the basis that the flexural rigidity is the same for all members

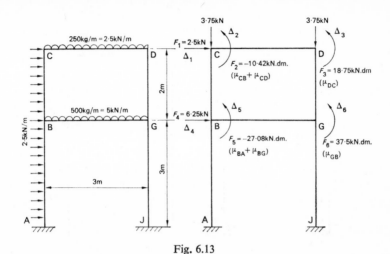

Fig. 6.13

with the exception of BG, which has a flexural rigidity of $2EI$. For the loading shown in Fig. 6.13 the final equations of equilibrium are as follows:

$$EI(0.0030\Delta_1 + 0.0150\Delta_2 + 0.0150\Delta_3 \\ -0.0030\Delta_4 + 0.0150\Delta_5 + 0.0150\Delta_6) = 2.50$$

$$EI(0.0150\Delta_1 + 0.3333\Delta_2 + 0.0667\Delta_3 \\ -0.0150\Delta_4 + 0.1000\Delta_5 + 0\Delta_6) = -10.42$$

$$EI(0.0150\Delta_1 + 0.0667\Delta_2 + 0.3333\Delta_3 \\ -0.0150\Delta_4 + 0\Delta_5 + 0.1000\Delta_6) = 18.75 \qquad (6.22)$$

$$EI(0.0030\Delta_1 - 0.0150\Delta_2 - 0.0150\Delta_3 \\ +0.0039\Delta_4 - 0.0083\Delta_5 - 0.0083\Delta_6) = 6.25$$

$$EI(0.0150\Delta_1 + 0.1000\Delta_2 + 0\Delta_3 \\ -0.0083\Delta_4 + 0.5000\Delta_5 + 0.1333\Delta_6) = -27.08$$

$$EI(0.0150\Delta_1 + 0\Delta_2 + 0.1000\Delta_3 \\ -0.0083\Delta_4 + 0.1333\Delta_5 + 0.6000\Delta_6) = 37.50$$

whence:

$$\Delta_1 = 15113/EI \text{ dm}; \qquad \Delta_2 = -96.2/EI \text{ rad};$$
$$\Delta_3 = -33.6/EI \text{ rad}; \qquad \Delta_4 = 12060/EI \text{ dm}; \qquad (6.23)$$
$$\Delta_5 = -219.0/EI \text{ rad}; \qquad \Delta_6 = -94.3/EI \text{ rad}.$$

where EI is in kN dm².

123

It is instructive to consider the solution of this problem by considering the load components separately having regard to symmetry as shown in Fig. 6.14. There are then two groups of simultaneous equations for solution instead of the six equations (6.22); one group of four and one of two equations.

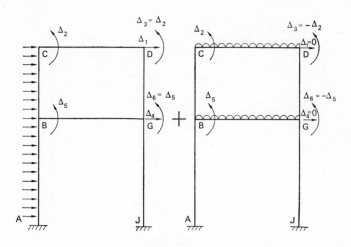

Fig. 6.14

(c) Although the uniform framework shown in Fig. 6.15 has only three degrees of freedom, namely, the rotational freedom at B and C and the sway freedom of the frame as a whole, and three redundants, inclination of a stanchion introduces some complication. Thus, while calculation of the stiffness coefficients in respect of rotation at B and C is readily accomplished, those in respect of sway are complicated by the fact that distortion of the beam BC is involved, as shown in Fig. 6.15(b). When sway occurs, B and C move sideways by the same amount, if second order small quantities are disregarded, and at the same time C rises by half this amount because, if ψ_1 is the angle of sway of the line AB ψ_2 is the angle of sway of CD and ψ_3 is the angle of rotation of BC:

$$h\psi_1 = h\psi_2; \qquad l\psi_3 = -\psi_2 h \cot \theta = -\psi_2 h/2 \qquad (6.24)$$

or

$$\psi_1 = \psi_2; \qquad \psi_3 = -\psi_1 h/2l \qquad (6.25)$$

The most difficult stiffness coefficient to calculate is b_{11} and for this purpose either the law of conservation of energy or the principle of virtual work is useful. Using the former and having regard to Fig. 6.1(c) from which it appears that the bending couples induced at the ends of a uniform encastré

124

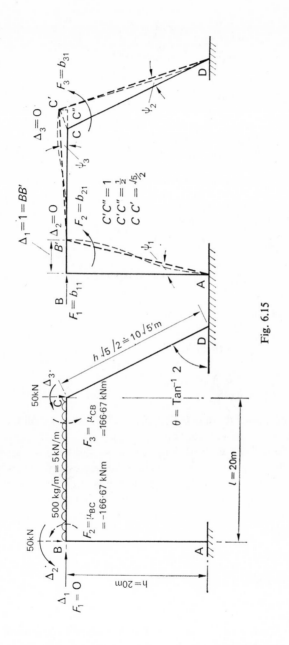

Fig. 6.15

member which is caused to sway through a small angle ψ are each of magnitude $(6EI/l^2) \times \psi \times l$:

$$\frac{1}{2} \times F_1 \times \Delta_1 = \frac{1}{2}\left[\frac{6EI}{h^2} \times \psi_1 \times h \times \psi_1 \times 2 + \frac{6EI}{(h\sqrt{5}/2)^2}\right.$$
$$\left. \times \psi_2 \times \frac{h\sqrt{5}}{2} \times \psi_2 \times 2 + \frac{6EI}{l^2} \times \psi_3 \times l \times \psi_3 \times 2\right] \quad (6.26)$$

When $\Delta_1 = 1$, $F_1 = b_{11}$ and by equation (6.25) $\psi_1 = \psi_2 = 1/h$; $\psi_3 = -\psi_1 h/2l$ $= -1/2l$, therefore:

$$b_{11} = 12EI\left(\frac{2 + \sqrt{5}}{h^3\sqrt{5}} + \frac{1}{4l^3}\right) = 0.0032EI \quad (6.27)$$

The remaining stiffness coefficients are readily determined as follows:

$$b_{12} = b_{21} = 6EI\left(\frac{1}{h^2} - \frac{1}{2l^2}\right) = 0.0075EI$$

(since $\psi_3 = -\psi_1/2$)

$$b_{13} = b_{31} = 6EI\left(\frac{1}{h^2\sqrt{5}} - \frac{1}{2l^2}\right) = 0.0059EI$$

$$b_{22} = 4EI\left(\frac{1}{h} + \frac{1}{l}\right) = 0.4000EI$$

$$b_{23} = b_{32} = \frac{2EI}{l} = 0.1000EI$$

$$b_{33} = 4EI\left(\frac{1}{l} + \frac{2}{h\sqrt{5}}\right) = 0.3780EI$$

$$(6.28)$$

For the loading shown it appears that $F_1 = 0$ but closer examination of the problem reveals that the vertical force of 50 kN at C induces sway of the framework. In fact, if the vertical deflection of C had been designated Δ_1 to define the sway, then $F_1 = -50$ kN. It is, however, necessary now to determine the value of F_1 applied horizontally at B, which is equivalent to the vertical force of 50 kN at C. This may be done by considering the equilibrium of the framework, transformed into a four-pin mechanism, subjected to F_1 at B and 50 kN vertically at C. By the principle of virtual work, then:

$$F_1 \times h \times \psi_1 = -50 \times l \times \psi_3 \quad (6.29)$$

whence, having regard to equation (6.25):

$$F_1 = 50/2 \text{ kN} \quad (6.30)$$

The value of F_1 for use in the final equations of equilibrium must, then, be $-50/2$ kN to cause the same sway as a downward load of 50 kN at C.

The final equations of equilibrium of the framework are as follows, taking into account equations (6.27) and (6.28), and putting $F_1 = -25$ kN; $F_2 = -166\cdot67$ kN m; $F_3 = 166\cdot67$ kN m

$$EI(0\cdot0032\Delta_1 + 0\cdot0075\Delta_2 + 0\cdot0059\Delta_3) = -25$$
$$EI(0\cdot0075\Delta_1 + 0\cdot4000\Delta_2 + 0\cdot1000\Delta_3) = -166\cdot67 \qquad (6.31)$$
$$EI(0\cdot0059\Delta_1 + 0\cdot1000\Delta_2 + 0\cdot3780\Delta_3) = 166\cdot67$$

whence:

$$\Delta_1 = -8170/EI \text{ m}; \qquad \Delta_2 = -434/EI \text{ rad};$$
$$\Delta_3 = 682/EI \text{ rad}. \qquad (6.32)$$

where EI is in kN m²

Single-bay frameworks of this kind, including pitched roof portals are analysed more easily, in general, by means of the compatibility approach as set out in Chapter 5. Thus, while the pitched roof portal framework whose stanchions are encastré in rigid foundations, has three redundants, it has five degrees of freedom (three in respect of joint rotation and two in respect of sway). By the equilibrium approach there are then five final equations for solution which, combined with the initial labour of calculating the stiffness coefficients of the framework and determining the loading referred to the joints, presents a formidable task.

6.7 Secondary effects in rigidly jointed trusses

Secondary effects in rigidly jointed trusses are the bending of the members which occurs when such a frame is loaded. Thus, the small changes in the angles between members which occur freely when a pin-jointed truss is loaded, are opposed in a rigidly jointed truss. As the term implies, secondary effects are usually small and can be estimated by assuming that the translational deflections of the joints of the truss are the same as if it were pin-jointed. This assumption is justified by experimental evidence to the extent that laboratory tests on small scale rigidly jointed trusses indicate that deflections under load are between 5 and 10 per cent less than the trusses with pin-joints.

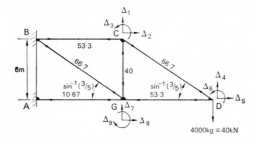

Fig. 6.16

The rigidly jointed truss with nine degrees of freedom and nine redundants shown in Fig. 6.16 is made of metal and each member has a cross sectional area of 140 cm² and a second moment of area, of 5330 cm⁴. In order to investigate the secondary effects it is first necessary to calculate the deflections of the joints due to the loading shown, assuming the joints are pinned. The axial or primary forces in the members due to the loading and calculated by elementary statics are shown in Fig. 6.16 and the consequent changes in length of the members are related to the deflections of the joints as follows, in metres:

$$e_{BC} = \Delta_2 = 5\cdot33 \times 8 \times 10^5/140E$$

$$e_{CG} = \Delta_1 - \Delta_7 = -4 \times 6 \times 10^5/140E$$

$$e_{BG} = -0\cdot6\Delta_7 + 0\cdot8\Delta_8 = 6\cdot67 \times 10 \times 10^5/140E$$

$$e_{GD} = \Delta_5 - \Delta_8 = -5\cdot33 \times 8 \times 10^5/140E \tag{6.33}$$

$$e_{AG} = \Delta_8 = -10\cdot67 \times 8 \times 10^5/140E$$

$$e_{CD} = 0\cdot6(\Delta_1 - \Delta_4) - 0\cdot8(\Delta_2 - \Delta_5) = 6\cdot67 \times 10 \times 10^5/140E$$

where Young's modulus E is in units of kN/m². By means of these relationships, the deflections of the joints in metres (Fig. 6.16) are:

$$\Delta_1 = -1\cdot794 \times 10^5/E; \qquad \Delta_2 = +0\cdot307 \times 10^5/E;$$

$$\Delta_4 = -4\cdot232 \times 10^5/E; \qquad \Delta_5 = -0\cdot922 \times 10^5/E; \tag{6.34}$$

$$\Delta_7 = -1\cdot621 \times 10^5/E; \qquad \Delta_8 = -0\cdot614 \times 10^5/E$$

The equations of equilibrium relating to the three degrees of freedom in respect of rotation of the joints of the rigidly jointed truss in terms of stiffness coefficients and deflections and rotations of the joints are as follows:

$$b_{31}\Delta_1 + b_{32}\Delta_2 + b_{33}\Delta_3 + b_{34}\Delta_4 + b_{35}\Delta_5 + b_{36}\Delta_6$$
$$+ b_{37}\Delta_7 + b_{38}\Delta_8 + b_{39}\Delta_9 = 0$$

$$b_{61}\Delta_1 + b_{62}\Delta_2 + b_{63}\Delta_3 + b_{64}\Delta_4 + b_{65}\Delta_5 + b_{66}\Delta_6$$
$$+ b_{67}\Delta_7 + b_{68}\Delta_8 + b_{69}\Delta_9 = 0 \tag{6.35}$$

$$b_{91}\Delta_1 + b_{92}\Delta_2 + b_{93}\Delta_3 + b_{94}\Delta_4 + b_{95}\Delta_5 + b_{96}\Delta_6$$
$$+ b_{97}\Delta_7 + b_{98}\Delta_8 + b_{99}\Delta_9 = 0$$

where the deflections Δ_1, Δ_2, Δ_4, Δ_5, Δ_7 and Δ_8 are given by equations (6.34) and Δ_3, Δ_6 and Δ_9, being the rotational deflections of C, D and G respectively, are unknown and can be found by means of these equations. The stiffness coefficients are obtained by considering unit translational and rotational displacements of the joints existing in turn, in the manner indicated in Fig. 6.17 for joint C. Thus, the relevant stiffness coefficients are given in Table 6.5.

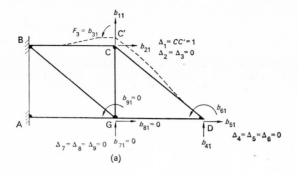

(a)

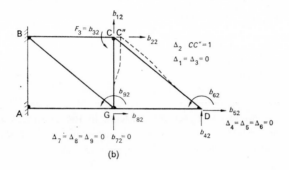

(b)

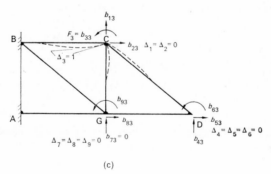

(c)

Fig. 6.17

Therefore, finally:

$$1\cdot567E\Delta_3 + 0\cdot200E\Delta_6 + 0\cdot333E\Delta_9 + 4\cdot839 \times 10^4 = 0$$
$$0\cdot200E\Delta_3 + 0\cdot900E\Delta_6 + 0\cdot250E\Delta_9 + 7\cdot083 \times 10^4 = 0 \qquad (6.36)$$
$$0\cdot333E\Delta_3 + 0\cdot250E\Delta_6 + 2\cdot067E\Delta_9 + 6\cdot496 \times 10^4 = 0$$

whence:

$$\Delta_3 = -3\cdot690 \times 10^4/E \text{ rad}; \qquad \Delta_6 = -3\cdot003 \times 10^4/E \text{ rad}; \qquad (6.37)$$
$$\Delta_9 = -2\cdot393 \times 10^4/E \text{ rad}.$$

5*

TABLE 6.5

$b\rightarrow$ $\downarrow$	1	2	3	4
3	$6EI\left(\dfrac{0\cdot8}{CD^2}-\dfrac{1}{BC^2}\right)$ $=-0\cdot0456EI$	$6EI\left(\dfrac{0\cdot6}{CD^2}+\dfrac{1}{CG^2}\right)$ $=0\cdot2028EI$	$4EI\left(\dfrac{1}{BC}+\dfrac{1}{CD}+\dfrac{1}{CG}\right)$ $=1\cdot5668EI$	$-6EI\left(\dfrac{0\cdot8}{CD^2}\right)$ $=-0\cdot0480EI$
6	$6EI\left(\dfrac{0\cdot8}{CD^2}\right)$ $=0\cdot0480EI$	$6EI\left(\dfrac{0\cdot6}{CD^2}\right)$ $=0\cdot0360EI$	$2EI\left(\dfrac{1}{CD}\right)$ $=0\cdot2000EI$	$-6EI\left(\dfrac{0\cdot8}{CD^2}+\dfrac{1}{DG^2}\right)$ $=-0\cdot1416EI$
9	0	$6EI\left(\dfrac{1}{CG^2}\right)$ $=0\cdot1667EI$	$2EI\left(\dfrac{1}{CG}\right)$ $=0\cdot3333EI$	$-6EI\left(\dfrac{1}{DG^2}\right)$ $=-0\cdot0936EI$

In order to calculate the bending moments in the members at the joints due to the secondary effects it is necessary to make use of equations (6.2); for example, the bending moment at G in member DG is:

$$M_{GD} = \frac{2EI}{DG}\left[\Delta_6 + 2\Delta_9 - \frac{3}{DG}(\Delta_4 - \Delta_7)\right]$$
$$= \frac{2\times5330}{8\times10^4}\left[-7\cdot789 + \frac{3\times26\cdot112}{8}\right] = 0\cdot270 \text{ kN m} \qquad (6.38)$$

that is, an anti-clockwise couple of $0\cdot270$ kN m is exerted on member DG at G by the other members which meet at G; where $\Delta_6 = \phi_D$; $\Delta_9 = \phi_G$ and $(\Delta_4 - \Delta_7)/DG = \psi_{DG}$.

6.8 Identification of the degrees of freedom of some typical complicated frameworks

(*a*) The fourteen degrees of freedom of a four-panel Vierendeel girder are indicated in Fig. 6.18 (the number of redundants is twelve). If the loading is

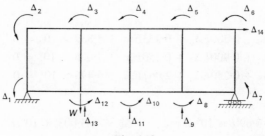

Fig. 6.18

	6	7	8	9
$-6EI\left(\dfrac{0\cdot6}{CD^2}\right)$	$2EI\left(\dfrac{1}{CD}\right)$		$-6EI\left(\dfrac{1}{CG^2}\right)$	$2EI\left(\dfrac{1}{CG}\right)$
$= -0\cdot0360EI$	$= 0\cdot2000EI$	0	$= -0\cdot1667EI$	$= 0\cdot3333EI$
$-6EI\left(\dfrac{0\cdot6}{CD^2}\right)$	$4EI\left(\dfrac{1}{CD}+\dfrac{1}{DG}\right)$	$6EI\left(\dfrac{1}{DG^2}\right)$		$2EI\left(\dfrac{1}{DG}\right)$
$= -0\cdot0360EI$	$= 0\cdot9000EI$	$= 0\cdot0938EI$	0	$= 0\cdot2500EI$
	$2EI\left(\dfrac{1}{DG}\right)$	$-6EI\left(\dfrac{0\cdot8}{BG^2}\right)$	$-6EI\left(\dfrac{0\cdot6}{BG^2}+\dfrac{1}{CG^2}\right)$	$4EI\left(\dfrac{1}{AG}+\dfrac{1}{BG}+\dfrac{1}{CG}+\dfrac{1}{DG}\right)$
	$= 0\cdot2500EI$	$= -0\cdot0480EI$	$= -0\cdot2028EI$	$= 2\cdot0667EI$

unsymmetrical, as shown, provided the frame is linear, the principle of super-position can be used to advantage in transforming the analysis into the analysis of the symmetrical and anti-symmetrical conditions shown in Fig. 6.19.

(*b*) The degrees of freedom of two unsymmetrical frameworks are illustrated in Figs. 6.20 and 6.21. In these instances the uses of the device of symmetry and anti-symmetry of loading is unlikely to be advantageous. It is important to note that pin-joints introduced at the feet of columns increase the number of degrees of freedom by one for each such joint and reduce the number of redundants by one for each such joint.

6.9 Effect of axial forces on the stiffness in bending of members of frameworks

This subject is outside the scope of this book and the following brief account is provided for the general guidance of the reader. The bending stiffness of structural members is reduced when they carry axial compressive loads and increased by axial tensile loads. It is the former which is sometimes critical to the load-carrying function of frameworks and requires careful investigation if the axial compressive force in any member is more than a small fraction of the parameter $P_e = \pi^2 EI/l^2$. This parameter is the well-known Euler critical load of the member as an ideal pin-ended strut, where l is its overall length (regardless of the nature of its terminal restraints in the framework) and EI is its flexural rigidity relevant to the plane of the framework. That is the Euler load if the member with pinned ends is prevented from bending out of the plane of the framework.

131

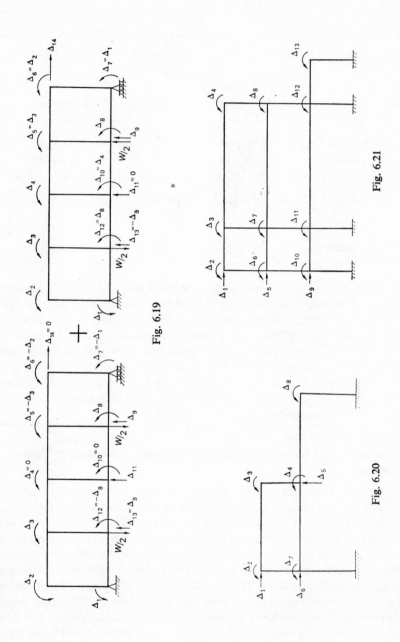

Fig. 6.19

Fig. 6.20

Fig. 6.21

The reduction in stiffness of axially loaded members may be taken into account by using what are commonly called s and c functions. For further details and values of these non-linear functions, which depend upon the ratio P/P_e (where P is the axial compressive load of the member) reference should

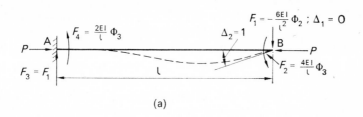

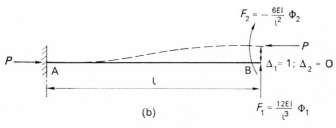

Fig. 6.22

be made to the work of Livesley (bibliography, ref. 11). The changes which are necessary in respect of Fig. 6.1(b) and (c) are, however, shown in Fig. 6.22, where the functions Φ are as follows:

$$\Phi_1 = s(1 + c)/6 - Pl^2/12EI; \qquad \Phi_2 = s(1 + c)/6;$$
$$\Phi_3 = s/4; \qquad \Phi_4 = sc/2 \tag{6.39}$$

In order to make use of these functions it is necessary first to analyse the framework in the manner described in the chapter, neglecting the axial loads on members. By the results of this analysis, the forces throughout the framework may be calculated using conditions of equilibrium, including the axial loads P on the members. These values of P provide a first approximation to enable a trial set of values of the ratios P/P_e to be calculated for the various members and thence trial values of s and c from graphs or tables (ref. 11). Having then calculated the quantities of equation (6.39) for each member in compression, a revised set of stiffness coefficients for the framework may be calculated on the basis of the data given in Fig. 6.22. Use of the revised stiffness coefficients for the purpose of a second analysis of the framework will serve to determine whether the deflections of the joints (and the values of P) are altered significantly in comparison with the results of the first analysis. If there is little change the process is complete; otherwise the process must be

133

repeated using revised values of P for determining the values of s and c, and so on. If the process does not give values of deflection which are convergent, frame instability is indicated. This is clearly a laborious process unless an automatic computer is used and the results of the second analysis are usually sufficient to indicate whether or not modification of the structure is desirable to ensure a sufficient margin of safety against instability.

A complete structure frequently consists of a system of orthogonal plane frameworks, the columns or stanchions of which are common. It is, therefore, necessary to consider each plane framework of the system. If a plane framework is to function in isolation it is important to investigate its lateral stability in addition to that in its own plane. Safety in this respect may usually be ensured by the provision of secondary bracing instead of increasing the sizes of main members. Moreover, in many structures the additional strength provided by decking, cladding or flooring is significant.

EXERCISES

1 Problems 2 and 3 of Chapter 5 by the methods of Chapter 6.

2 A uniform rigidly jointed portal framework ABCD has vertical stanchions AB and CD of length 10 m and beam BC of length 20 m. Calculate the bending couples at B and C due to a load of 1000 kg acting vertically on BC at a point 5 m from B.

Ans: 17·28 kN m; 12·48 kN m

3 A uniform rigidly jointed framework ABCD consists of a vertical stanchion AB of length l, a horizontal beam BC of length l and an inclined stanchion CD of length $l\sqrt{2}$ such that the foundations A and D are at the same level and the distance between them is $2l$. If the stanchions are encastré in rigid foundations at A and D obtain expressions for the deflections of joints B and C due to a uniformly distributed load of intensity w per unit length of the beam AB. The expressions should be in terms of w, l and EI.

Ans: $-0{\cdot}0146wl^4/EI$; $-0{\cdot}0136wl^3/EI$; $+0{\cdot}0126wl^3/EI$

4 A rigidly jointed plane framework ABC in the form of an isosceles triangle is such that AB = 1·2 m and AC = BC = $1{\cdot}2/\sqrt{3}$ m. It is supported so that its base AB is horizontal and its apex C is below AB. Calculate the secondary bending moments at C due to a load of 4000 kg suspended from that point if each member of the framework has a cross-sectional area of 1000 mm² and a second moment of area of 400 000 mm⁴.

Ans: 0·275 kN m

5 A simply supported rigidly jointed open bay (Vierendeel) type of bridge girder consists of four bays of equal length l. The two end bays are triangular in shape and the two inner bays are square. If the flexural rigidity of all members is EI show that the deflection at mid span due to a concentrated load F acting there is $0{\cdot}022Fl^3/EI$. Axial deformation of the members should be neglected.

6 Solve the problem of the grid described in para. 8.3 by the equilibrium approach, neglecting the effect of torsion of members.

Ans: $\Delta_{max} = 0{\cdot}0066Fl^3/EI$

7 Describe the use of the equilibrium approach for the framework of problem 2 if there is a pin-joint at **B**.

7

Some uses of the reciprocal theorem: model analysis

7.1 As noted in para. 2.8 the reciprocal theorem of linear systems, specifies that the deflection in the line i of an elastic structure due to the application of unit force in the line j is equal to the deflection in the line j when unit force is applied in the line i. The lines i and j generally pass through different points on the structure though they may pass through the same point. It is manifest when the flexibility coefficients of linear structures are calculated, since then it is found that $a_{ij} = a_{ji}$ as shown in Chapter 2. It is also manifest when the stiffness coefficients are calculated.

For the purpose of structural analysis the reciprocal theorem provides useful devices for the construction of influence lines for deflections and forces of frameworks whose elasticity is linear.

7.2 Influence lines for deflection by the reciprocal theorem

For the purpose of illustrating this use of reciprocal theorem it is sufficient to consider a simply supported beam with linear elasticity. Thus, if the influence line for the deflection of any point P of the beam shown in Fig. 7.1 is required (that is, the curve whose ordinates represent the deflection of P as a concentrated unit load traverses the beam), by the reciprocal theorem it is merely necessary to consider the deflected shape of the beam due to unit load at P. The reason for this is that the deflection at any other point Q of the beam due to unit load at P is:

$$\Delta_Q = a_{QP} . 1 \tag{7.1}$$

where a_{QP} is the relevant flexibility coefficient. Since this is equal to the deflection of P due to unit load at Q, i.e.:

$$\Delta_P = a_{PQ} . 1 \tag{7.2}$$

it follows that the deformed shape of the beam caused by unit load at P represents the variation of $a_{QP} = a_{PQ}$ over the length of the beam and is the influence line for the deflection of P. By similar reasoning the influence line for the deflection of any point of any linearly elastic structure, in a given direction, is represented by the deformed shape of the structure due to unit load applied in the specified direction at that point.

A convenient means of using this principle to practical advantage is afforded by scale models.[1] Such models need not be to scale in every detail; for plane

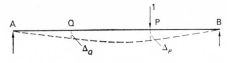

Fig. 7.1

frameworks it is merely necessary that they are made of material which obeys Hooke's law of linear elasticity, to a chosen layout scale. Then, for portal frameworks whose members deform primarily in bending, it it sufficient for the ratios of the second moments of area of the members to be the same as in the actual framework. The shape of the required influence line to scale can be obtained by applying a force to the model at the point in question, in the specified direction. The scale factor for the ordinates of the influence line so obtained can be found either by scaling the force applied to the model or by calculating the deflection of the actual framework due to unit load applied at the point for which the influence line is required.

7.3 Influence lines for forces by the reciprocal theorem: Müller-Breslau's principle

A cantilever with a rigid prop at its 'free' end, as one of the simplest statically indeterminate systems, is suitable for demonstrating this use of the reciprocal theorem. In order to obtain the influence line for the force exerted by the prop, suppose first of all that unit concentrated load acts at any point Q of the span, as shown in Fig. 7.2(a). If the prop is absent the deflection of the end of the cantilever due to this load is:

$$\Delta_\mathrm{P} = a_\mathrm{PQ}.1 \tag{7.3}$$

so that the force which the prop must exert in restoring zero deflection at this point is:

$$R_\mathrm{P} = \frac{\Delta_\mathrm{P}}{a_\mathrm{PP}} \tag{7.4}$$

where the flexibility coefficients a_PQ and a_PP refer to the cantilever. Therefore, by equations (7.3) and (7.4):

$$R_\mathrm{P} = \frac{a_\mathrm{PQ}}{a_\mathrm{PP}} \tag{7.5}$$

[1] See bibliography, refs. 1 and 13.

Now the ratio a_{PQ}/a_{PP} can be obtained by considering an arbitrary small displacement Δ'_P of the end of the unloaded cantilever due to an arbitrary force R'_P, as shown in Fig. 7.2(b) since:

$$\Delta'_P = a_{PP}R'_P \qquad (7.6)$$

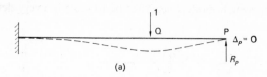

(a)

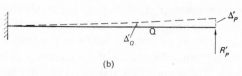

(b)

Fig. 7.2

while the resulting deflection of any other point Q is:

$$\Delta'_Q = a_{QP}R'_P \qquad (7.7)$$

so that:

$$\frac{\Delta'_Q}{\Delta'_P} = \frac{a_{QP}}{a_{PP}} = \frac{a_{PQ}}{a_{PP}} \qquad (7.8)$$

Therefore, by equation (7.5):

$$R_P = \frac{\Delta'_Q}{\Delta'_P} \qquad (7.9)$$

The significance of this result is that the deflection curve of the cantilever due to an arbitrary small displacement of P represents to scale the influence line for the load on the prop at P. This is in accordance with Müller-Breslau's principle [1] that the influence line for the force in a member or upon a support of a linear statically indeterminate framework is represented to scale by the change in shape of the framework due to a small displacement within the member or at the support. For the purpose of using the principle for the influence line for the bending moment at any point, the small displacement introduced there must be of the angular kind. It can be shown by virtual work that Müller-Breslau's principle also applies to statically determinate systems which are not subject to gross distortion under load.

Müller-Breslau's principle would be of very little practical value without scale model techniques. The procedure prescribed by the principle can be applied physically to a scale model for the purpose of obtaining influence lines

[1] See appendix 1.

138

to scale and affords an effective method of 'model analysis' of frameworks. Such models must be made of material with linear elasticity to a definite length scale. Thus, if a model of the propped cantilever is made s times smaller than the actual, a small displacement $(\Delta_P)_m$ at P corresponds to a small displacement $\Delta_P = s(\Delta_P)_m$ at P of the actual system. Similarly, the displacement of any other point of the model Q may be multiplied by the scale factor s to obtain the corresponding displacement of the point Q of the actual cantilever due to the displacement of P of $s(\Delta_P)_m$. Also the deformed shape of the model represents the influence line for the load on the prop of the actual system to scale. Therefore, with reference to equations (7.8) and (7.9):

$$\frac{a_{PQ}}{a_{PP}} = \frac{\Delta_Q'}{\Delta_P'} = \frac{s(\Delta_Q')_m}{s(\Delta_P')_m} = \frac{(\Delta_Q')_m}{(\Delta_P')_m} \qquad (7.10)$$

so that:

$$R_P = \frac{(\Delta_Q')_m}{(\Delta_P')_m} \qquad (7.11)$$

and the scale factor does not appear in the final result obtained by the model in respect of influence lines for forces, because the ratios of model displacements of the linear kind are identical to the ratios of corresponding displacements of the actual structure.

It is relatively easy to construct suitable models of frameworks whose members deform primarily in bending, such as portals, because then it is merely necessary for the ratios of the second moments of area of the various members to be correct. The actual scale factor in respect of second moment of area is immaterial and models can be cut from, say, sheet celluloid, which obeys Hooke's law. Beggs pioneered the use of this kind of model (see Appendix I).

7.4 Example of model analysis

The linearly elastic portal framework shown in Fig. 7.3 has encastré stanchion feet and the second moments of area of AB, BC and CD are I, $2I$ and I,

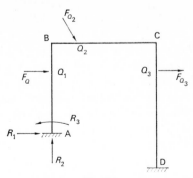

Fig. 7.3

respectively. In order to obtain the influence lines for the redundants, chosen to be the reactions R_1, R_2 and R_3 at the foot A, a scale model may be used. The model must be made of material which has linear elasticity in accordance with Hooke's law (e.g. it can be cut from sheet Xylonite celluloid), to a layout scale factor s and the ratios of the second moments of area of the model

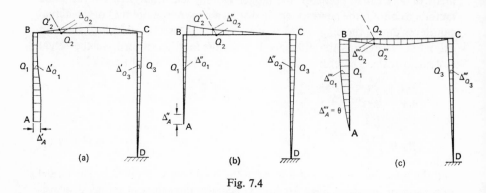

Fig. 7.4

members AB, BC and CD must be $1:2:1$ (i.e. the depths of the members cut from sheet must be in the ratios $1:^3\sqrt{2}:1$). The required influence lines are found by subjecting the model, mounted to reproduce the encastré conditions at A and D, to small displacements horizontally (for the influence line for R_1), vertically (for R_2), and rotationally (for R_3) at A, in turn, and recording the resulting changes in shape of the model. It is important for each displacement to be applied at A separately without movement in any other direction.

Suppose the influence lines so obtained are as shown in Fig. 7.4 and that it is desired to determine the magnitudes of the reactions at A caused by the loading shown in Fig. 7.3. Then using subscripts m to denote that the displacements are obtained from the model:

$$R_1 = -\frac{(\Delta'_{Q1})_m}{(\Delta'_A)_m} F_{Q1} + \frac{(\Delta'_{Q2})_m}{(\Delta'_A)_m} F_{Q2} - \frac{(\Delta'_{Q3})_m}{(\Delta'_A)_m} F_{Q3}$$

$$R_2 = -\frac{(\Delta''_{Q1})_m}{(\Delta''_A)_m} F_{Q1} + \frac{(\Delta''_{Q2})_m}{(\Delta''_A)_m} F_{Q2} - \frac{(\Delta''_{Q3})_m}{(\Delta''_A)_m} F_{Q3}$$

(7.12)

which are independent of the scale of the model. For R_3, however, the scale of the model enters into the calculations and for this reason it is desirable to refer the model displacements to the corresponding values for the actual structure. Thus, if the foot A of the actual framework were rotated through θ radians the resulting deflections would be s times those of the model when its foot A is rotated through the same angle. Using the equivalent full-scale influence line ordinates then to obtain R_3 gives:

$$R_3 = \frac{s(\Delta'''_{Q1})_m}{\theta} F_{Q1} - \frac{s(\Delta'''_{Q2})_m}{\theta} F_{Q2} + \frac{s(\Delta'''_{Q3})_m}{\theta} F_{Q3}$$

(7.13)

since θ is $\Delta_A''' = (\Delta_A''')_m$.

Again, for a uniformly distributed loading of intensity w over, say, CD, acting to the right the corresponding values of the reactions at A are:

$$-R_1 = \int_C^D \frac{(\Delta_{Qx}')_m}{(\Delta_A')_m} ws\, dx$$

$$-R_2 = \int_C^D \frac{(\Delta_{Qx}'')_m}{(\Delta_A'')_m} ws\, dx \qquad (7.14)$$

$$R_3 = s \int_C^D \frac{(\Delta_{Qx}''')_m}{\theta} ws\, dx$$

where distance x along CD refers to the model, so that if a_1, a_2 and a_3 are the areas enclosed by the relevant portions of the influence lines of the model respectively:

$$-R_1 = \frac{sa_1}{(\Delta_A')_m} w$$

$$-R_2 = \frac{sa_2}{(\Delta_A'')_m} w \qquad (7.15)$$

$$R_3 = \frac{s^2 a_3}{\theta} w$$

and for practical purposes it is sufficiently accurate to assume that the influence lines are straight between measured ordinates.

The influence line for the bending moment at a point within a member can be obtained similarly by cutting the model at the point and applying an angular displacement, as indicated in Fig. 7.5. The required bending moment due to particular loading is then obtained from the influence line ordinates in a manner similar to that used for finding R_3.

It is particularly important to measure the influence line ordinates correctly, as, for example, in Fig. 7.4 with respect to the line of F_{Q2}. Accuracy can also be improved by using positive and negative displacements, as shown in Figs. 7.7 and 7.8.

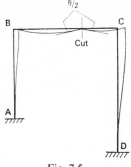

Fig. 7.5

Use of scale models for the analysis of frameworks and arches is always worth considering as an alternative to manual computation, especially for frameworks of simple form whose members are of non-uniform section for reasons of economy. Accuracy of model analysis tends to lie between 5 and 10 per cent in relation to values calculated exactly on the basis of the same assumptions as those used in constructing the model. The latter figure can be obtained by using high-grade cardboard for models of simple frameworks. This is particularly convenient for the casual user of the technique and the reader is recommended to verify it for himself. It is, however, possible to give here only some essential features of model analysis and for a more detailed treatment of the subject reference may be made to specialist works on the subject (e.g. bibliography, ref. 13.)

7.5 Betti's theorem[1]

The form of the reciprocal theorem derived by Betti and which is known as Betti's theorem is as follows:

$$F_1\Delta_1' + F_2\Delta_2' + \cdots + F_n\Delta_n' = F_1'\Delta_1 + F_2'\Delta_2 + \cdots + F_N'\Delta_N \tag{7.16}$$

where $F_1, F_2, \ldots, F_n$, and $F_1', F_2', \ldots, F_N'$ are systems of forces or couples applied separately to a linear elastic system and $\Delta_1, \Delta_2, \ldots, \Delta_N$ are the deflections produced by the former in the directions of the lines of action of the latter, respectively, while $\Delta_1', \Delta_2', \ldots, \Delta_n'$ are the deflections produced by the latter in

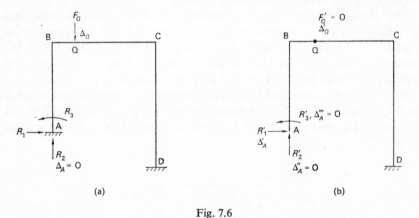

(a) (b)

Fig. 7.6

the directions of the lines of action of the former, respectively. This theorem, which was justified in an elegant manner by virtual work (in a manner similar to that of para. 2.8) can be applied directly for the purpose of verifying Müller-Breslau's principle with respect to linear statically indeterminate structures. For example, with reference to the portal framework shown in

[1] See bibliography, ref. 19.

Fig. 7.3 by considering the two systems of forces shown in Fig. 7.6 and applying Betti's theorem:

$$F_Q\Delta_Q' + R_1\Delta_A' + R_2.0 + R_3.0 = R_1'.0 + R_2'.0 + R_3'.0 \qquad (7.17)$$

If, therefore $F_Q = 1$, i.e. upward since the Cartesian convention is essential to Betti's theorem:

$$R_1 = -\frac{\Delta_Q'}{\Delta_A'} \qquad (7.18)$$

Thus, the deformed shape of the framework associated with an arbitrary displacement Δ_A' at A produced by the system of forces R_1', R_2' and R_3' represents the influence line for reaction R_1.

An interesting additional example of the use of Betti's theorem concerns a metal vessel subjected to a concentrated load F which causes a change in its internal volume of v determined by measuring the quantity of liquid displaced from the vessel at atmospheric pressure. It is required to find the deflection Δ' at the point of application of the load which is caused by subjecting the vessel to an internal pressure of p' above atmospheric pressure. By Betti's theorem:

$$F\Delta' = p'v \qquad (7.19)$$

whence:

$$\Delta' = \frac{p'v}{F} \qquad (7.20)$$

In this instance, application of the simple form of the reciprocal theorem tends to be tedious while use of Betti's form of the theorem provides a rapid and elegant solution.

EXERCISES

1 A steel spline model of an elastic arch is shown in Fig. 7.7. If positive and negative displacements of 1 cm of A, as shown, cause a total displacement of 1 cm at Q in the direction of the line of action of a force F_Q applied to the actual arch, find the abutment thrust at A when $F_Q = 100$ kN.

Ans: 50 kN

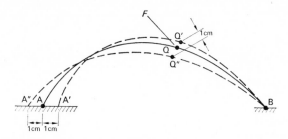

Fig. 7.7

2 A model of a rigidly jointed portal framework to a scale of 1:20 is shown in Fig. 7.8. Positive and negative rotational displacements of D of 0·1 rad, as shown, cause total horizontal movements of points 15 cm, 30 cm, 45 cm and 60 cm (C) above D of 0·96 cm, 1·32 cm, 1·78 cm and 2·29 cm, respectively.

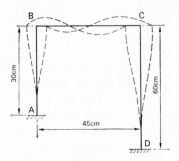

Fig. 7.8

Calculate the numerical value of the bending moment at D of the actual framework, (a) for a concentrated load of 1000 kg applied horizontally 9·1 m above D, and (b) for a uniformly distributed load of 330 kg/m applied over CD (the influence line may be assumed to consist of straight lines between each of the ordinates given).

Ans: 17·8 kN m; 52 kN m

3 For the framework shown in Fig. 7.9, when the support at C is removed and a force of 10 kN is applied vertically there the deflection of B or D is 0·64 cm and that of C is 2 cm. Construct the influence line for the reaction at C with reference to loading applied to the lower boom.

Ans: Ordinates 0; 0·32; 1·0; 0·32; 0

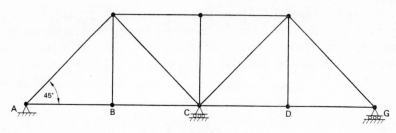

Fig. 7.9

144

8

Approximate solutions by energy methods

8.1 For certain kinds of elastic structure of some complexity it is possible to use energy principles to achieve rapid analysis which, though approximate, is nevertheless within bounds which are acceptable for engineering design. Included among such structures are beams with complicated loading, beams on elastic supports, suspension bridge systems and grid structures. The conservation of energy method or the principle of stationary total potential energy (para. 1.4) is perhaps the most useful for this kind of approximate analysis but the principle of stationary complementary energy can be useful also. The former can provide a lower bound for deflections at points of application of loads and an upper bound for internal force or bending moment while the latter can provide a lower bound for internal force or bending moment and an upper bound for deflection. Thus, the correct solution will usually be between those of the two energy principles.

8.2 The conservation of energy method

One of the simplest applications of this method is appropriate to illustrate its essential features. If a uniform simply supported beam of span l and flexural rigidity EI is subjected to a system of concentrated loads $F_1, F_2, \ldots, F_n$ over its span and it is desired to determine the form of its deflection approximately, it is first necessary to select a deflection function which satisfies the geometrical boundary conditions. In this instance the boundary conditions are merely that the deflection is zero at each end of the beam. If, moreover, the chosen function has only one unknown parameter, for example, if $\Delta = A \sin (\pi/l)x$ is assumed as satisfying the boundary conditions the procedure is very simple as follows:

By the law of conservation of energy for a state of equilibrium:

$$\delta \text{ (work done by the loading)} = \delta \text{ (strain energy)}$$

i.e.
$$\delta W = \delta U \tag{8.1}$$

for a small variation of the loading.

Principles of structural analysis

Now:

$$\delta W = \sum_{i}^{n} F_i \, \delta \Delta_i = \delta A \sum_{i}^{n} F_i \sin \frac{\pi}{l} x_i \tag{8.2}$$

if second order small quantities are neglected and where x_i locates the position of the ith component of the loading and is measured from the left-hand end of the beam. Also:

$$\delta U = \delta \left[\frac{EI}{2} \int_0^l \left(\frac{d^2 \Delta}{dx^2} \right)^2 dx \right] = EI \int_0^l \left(\frac{d^2 \Delta}{dx^2} \right) \delta \left(\frac{d^2 \Delta}{dx^2} \right) dx \tag{8.3}$$

with respect to the parameter (or parameters) of deflection, whence, substituting $\Delta = A \sin (\pi/l)x$:

$$\delta U = \frac{\pi^4 EIA \, \delta A}{l^4} \int_0^l \sin^2 \frac{\pi}{l} x \, dx = \frac{\pi^4 EIA \, \delta A}{2l^3} \tag{8.4}$$

By equation (8.1), then:

$$\delta A \sum_{i}^{n} F_i \sin \frac{\pi}{l} x_i = \frac{\pi^4 EIA \, \delta A}{2l^3}$$

or

$$A = \frac{2l^3}{\pi^4 EI} \sum_{i}^{n} F_i \sin \frac{\pi}{l} x_i \tag{8.5}$$

If the loading consists of only a force F at the centre of the span:

$$A = \frac{2Fl^3}{\pi^4 EI} = \frac{Fl^3}{48 \cdot 65 EI} \tag{8.6}$$

and represents the mid span deflection, the precise value of which by the simple theory of flexure is $Fl^3/48EI$. The degree of approximation with respect to deflection at this point is clearly very small therefore.

Again, if the loading is uniformly distributed of intensity w per unit length of the beam, then:

$$A = \frac{2l^3 w}{\pi^4 EI} \int_0^l \sin \frac{\pi}{l} x \, dx \tag{8.7}$$

or

$$A = \frac{4wl^4}{\pi^5 EI} = \frac{4wl^4}{306 EI} \tag{8.8}$$

which is almost identical with the value of $5wl^4/384EI$ which is provided by the simple theory of flexure.

Application of the method to these two types of loading for which the maximum deflection of a uniform simply supported beam is well known, serves to indicate the effectiveness. In both instances the approximate value is slightly less than the correct value on the basis of the assumption of the

146

simple theory of flexure and this feature is characteristic of the method. The explanation lies in the implication of the introduction of additional restraint in using an approximation for the distribution of deflection. Now additional restraint assists the beam in resisting loads so that, as might be expected, the approximate value of the maximum deflection is less than the correct value in each instance.

8.3 Application of the conservation of energy method to the analysis of statically indeterminate systems

The simple procedure described in para. 8.2 above may be used for statically indeterminate systems. For example, to analyse the loaded continuous structure with six elastic supports illustrated in Fig. 8.1 which has therefore six

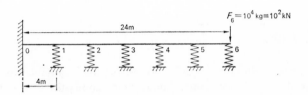

Fig. 8.1

redundants and twelve degrees of freedom (though strictly only six need be considered for loading of the kind shown), it is sufficient to assume that the distribution of downward deflection of the beam is represented by:

$$\Delta = A\left(1 - \cos\frac{\pi}{2l}x\right) \tag{8.9}$$

where l is the overall length of the beam and x is measured from the origin at the fixed end.

Now:

$$\delta U = b\sum_{}^{6}\Delta_i\,\delta\Delta_i + EI\int_0^l \left(\frac{d^2\Delta}{dx^2}\right)\delta\left(\frac{d^2\Delta}{dx^2}\right)dx \tag{8.10}$$

where b is the stiffness of each elastic support. Also:

$$\delta W = F_6\,\delta\Delta_6 \tag{8.11}$$

Substituting from equation (8.9) in equations (8.10) and (8.11) gives:

$$\delta U = bA\,\delta A\sum_{}^{6}\left(1 - \cos\frac{\pi}{2l}x_i\right)^2 + \frac{\pi^4 EIA\,\delta A}{(2l)^4}\int_0^l \cos^2\frac{\pi}{2l}x\,dx$$

$$= bA\,\delta A\sum_{}^{6}\left(1 - \cos\frac{\pi}{2l}x_i\right)^2 + \frac{\pi^4 EIA\,\delta A}{32l^3} \tag{8.12}$$

and

$$\delta W = F_6 \, \delta A \tag{8.13}$$

whence, by conservation of energy:

$$F_6 = bA \sum_{}^{6} \left(1 - \cos \frac{\pi}{2l} x_i\right)^2 + \frac{\pi^4 EIA}{32l^3} \tag{8.14}$$

or

$$A = \frac{F_6}{b \sum_{}^{6} \left(1 - \cos \frac{\pi}{2l} x_i\right)^2 + \frac{\pi^4 EI}{32l^3}} \tag{8.15}$$

Substituting now $F_6 = 10^2$ kN; $EI = 1\cdot5 \times 10^6$ kN m^2; $b = 10^3$ kN/m; $l = 24$ m; and putting $x_1 = 4$ m; $x_2 = 8$ m; ...; $x_6 = 24$ m; gives:

$$A = \frac{10^5}{1\cdot9048 \times 10^6 + \dfrac{1\cdot5 \times \pi^4 \times 10^9}{32 \times 24^3}} = 0\cdot045 \text{ m} \tag{8.16}$$

while the correct value for the deflection at the load, obtained by setting up and solving the six simultaneous equations of compatibility for the redundants, is 0·047 m. Having found A, the reaction R_i of any one elastic support may be obtained with the aid of equation (8.9) since $R_i = b\Delta_i$.

Yet another example of this method is afforded by the analysis of a grid loaded as shown in Fig. 8.2. In this instance all of the members of the grid are

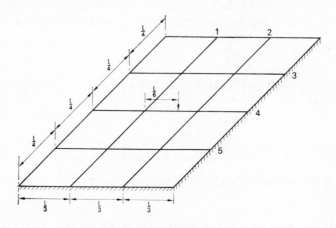

Fig. 8.2

of the same section and simply supported at their ends. It is assumed that the orthogonal members are connected together at the intersections and that torsion is negligible (these assumptions are made merely for reasons of simplicity and are not essential to the method). Because this is a symmetrical system subjected to a central concentrated load F, the distribution of deflection of each member may be represented approximately by a simple sine function as satisfying the geometrical boundary conditions. Thus:

for members 1 and 2: $\Delta_1 = \Delta_2 = A_1 \sin \dfrac{\pi}{l} x$ by symmetry;

for members 3 and 5: $\Delta_3 = \Delta_5 = A_3 \sin \dfrac{\pi}{l} y$ by symmetry;   (8.17)

for member 4: $\Delta_4 = A_4 \sin \dfrac{\pi}{l} y$

For compatibility of deflection at the intersections:

at $x = l/4$; $y = l/3$:

$$\Delta_1 = \Delta_3$$

therefore:

$$A_3 = A_1 \sqrt{\frac{2}{3}}$$

at $x = l/2$; $y = l/3$:

$$\Delta_1 = \Delta_4$$ (8.18)

therefore:

$$A_4 = A_1 \frac{2}{\sqrt{3}}$$

Thus, the deflections of the members are given by:

$$\Delta_1 = \Delta_2 = A_1 \sin \frac{\pi}{l} x$$

$$\Delta_3 = \Delta_5 = A_1 \sqrt{\frac{2}{3}} \sin \frac{\pi}{l} y$$ (8.19)

$$\Delta_4 = A_1 \frac{2}{\sqrt{3}} \sin \frac{\pi}{l} y$$

Now, by the law of conservation of energy $\delta W = \delta U$, where:

$$\delta U = 2EI \int_0^l \left(\frac{d^2\Delta_1}{dx^2}\right) \delta\left(\frac{d^2\Delta_1}{dx^2}\right) dx + 2EI \int_0^l \left(\frac{d^2\Delta_3}{dy^2}\right) \delta\left(\frac{d^2\Delta_3}{dy^2}\right) dy$$
$$+ EI \int_0^l \left(\frac{d^2\Delta_4}{dy^2}\right) \delta\left(\frac{d^2\Delta_4}{dy^2}\right) dy \quad (8.20)$$

and

$$\delta W = F \, \delta\Delta_{4\,max}$$ (8.21)

149

Substituting from equations (8.19) for Δ_1, Δ_3 and Δ_4, then gives:

$$\delta U = \frac{2\pi^4 EIA_1 \delta A_1}{l^4} \int_0^l \sin^2 \frac{\pi}{l} x \, dx + \frac{4\pi^4 EIA_1 \delta A_1}{3l^4} \int_0^l \sin^2 \frac{\pi}{l} y \, dy$$

$$+ \frac{4\pi^4 EIA_1 \delta A_1}{3l^4} \int_0^l \sin^2 \frac{\pi}{l} y \, dy \quad (8.22)$$

$$= \frac{7\pi^4 EIA_1 \, \delta A_1}{3l^3}$$

and

$$\delta W = \frac{2}{\sqrt{3}} F \, \delta A_1 \tag{8.23}$$

whence, by equating δU and δW:

$$A_1 = \frac{2\sqrt{3} \, Fl^3}{7\pi^4 EI} \tag{8.24}$$

so that the deflection at the load is:

$$\Delta = \frac{2}{\sqrt{3}} A_1 = \frac{4Fl^3}{7\pi^4 EI} = 0.0059 \frac{Fl^3}{EI} \tag{8.25}$$

This result should be compared with the correct value of $0.0066Fl^3/EI$, obtained either by the equilibrium approach or the compatibility approach. It will be noted that it is some 10 per cent less than the correct value but is probably closer to the result which is obtained when the effect of the torsional stiffness of rigidly connected members is taken into account.

In order to improve the accuracy of an approximate solution the distribution of deformation may be represented by a series with several undetermined parameters. Thus, for the system shown in Fig. 8.1 and considered above the deflection of the beam may be represented as follows:

$$\Delta = A_1 x^2 + A_2 x^3 \tag{8.26}$$

as satisfying the geometrical boundary conditions that $\Delta = d\Delta/dx = 0$ at $x = 0$. (Alternatively, a trigonometrical series could be used.)

Thus:

$$\frac{d^2\Delta}{dx^2} = 2A_1 + 6A_2 x \tag{8.27}$$

$$\delta\left(\frac{d^2\Delta}{dx^2}\right) = 2\delta A_1 + 6x \, \delta A_2 \tag{8.28}$$

and

$$\delta\Delta = x^2 \, \delta A_1 + x^3 \, \delta A_2 \tag{8.29}$$

Therefore, with reference to equations (8.10) and (8.11):

$$\delta U = 4EI \int_0^l (A_1 + 3A_2x)(\delta A_1 + 3\delta A_2x)\, dx + b\sum_{}^{6}(A_1x_i^2 + A_2x_i^3)$$

$$\times (x_i^2\, \delta A_1 + x_i^3\, \delta A_2) = F_6l^2(\delta A_1 + l\delta A_2) \quad (8.30)$$

whence:

$$\frac{\partial U}{\partial A_1} = 4EI \int_0^l (A_1 + 3A_2x)\, dx + b\sum_{}^{6} x_i^2(A_1x_i^2 + A_2x_i^3) = F_6l^2$$

$$\frac{\partial U}{\partial A_2} = 12EI \int_0^l (A_1 + 3A_2x)x\, dx + b\sum_{}^{6} x_i^3(A_1x_i^2 + A_2x_i^3) = F_6l^3$$

$$(8.31)$$

being two simultaneous equations of equilibrium in the two unknown parameters A_1 and A_2, the terms on their right-hand sides representing the loading referred to those parameters. Substituting the values $EI = 1\cdot5 \times 10^6$ kN m^2; $b = 10^3$ kN/m; $l = 24$ m; $F_6 = 10^2$ kN and solving gives:

$$A_1 = 8\cdot45 \times 10^{-5}\, \text{m}^{-1}; \qquad A_2 = 2\cdot075 \times 10^{-7}\, \text{m}^{-2} \quad (8.32)$$

whence:

$$R_1 = 1\cdot3\, \text{kN}; \qquad R_2 = 5\cdot3\, \text{kN}; \qquad R_3 = 11\cdot84\, \text{kN};$$
$$R_4 = 20\cdot8\, \text{kN}; \qquad R_5 = 32\cdot1\, \text{kN}; \qquad R_6 = 45\cdot8\, \text{kN} \quad (8.33)$$

Comparison with the result of equation (8.16) indicates that the maximum deflection (represented by R_6) is now closer to the correct value (bibliography, ref. 19) of 0·047 m, though it still represents a lower bound.

Calculation of the maximum bending moment (which is located at R_3 where the shearing force changes sign) by elementary statics gives a value of 310·4 kN m, which may be compared with the correct value of 305·7 kN m. That is, the maximum bending moment calculated from the results of the approximate analysis by the conservation of energy method represents an upper bound. It should be noted that having determined the distribution of bending moment on the basis of the conservation of energy method, its form may be used to determine a suitable series for use with the complementary energy method (para. 8.6).

8.4 The conservation of energy method for the approximate analysis of suspension bridge structures

In recent years it has been demonstrated, notably by Pugsley (bibliography, ref. 22) that the behaviour of a uniformly loaded cable in respect of additional loading which causes only small deflection, may be described in the manner of a linearly elastic structure. There is, however, a limitation in that the ratio of span to dip of the cable should not be less than about 10.

Now a suspension bridge usually fulfils the necessary conditions. The cables are uniformly loaded for practical purposes by the dead load of the deck in addition to their own weight and the stiffness which the cables thus acquire due to gravity is such that their load/deflection relationship for additional or live loading (which is small in comparison with the total dead load) is linear or very nearly so. The structure may, therefore, be considered as a linearly elastic structure representing the dead-loaded cable system, supporting a linearly elastic deck by hangers, as shown in Fig. 8.3.

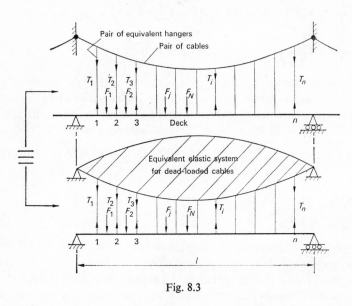

Fig. 8.3

The axial deformation of the hangers of a suspension bridge is usually negligible in comparison with the deflections of cable system and deck (which are, therefore, identical). Thus, it is easy to simplify a given structure further for purposes of analysis by reducing the number of hangers (or pairs of hangers) to, say, a dozen equivalent hangers. Such simplification implies, however, a reduction of stiffness of the structure which in approximate conservation or potential energy solutions tends to offset the increase in stiffness implied in approximation of the distribution of deflection.

The conservation of energy equation relating to a small variation of the live loads on the deck of the structure is as follows:

$$\delta U = \delta U_{\mathrm{C}} + \delta U_{\mathrm{D}} = \sum_{}^{N} F_j \, \delta \Delta_j \qquad (8.34)$$

where $\delta U_{\mathrm{C}} = \sum^{n} T_i \, \delta \Delta_i$, being the variation of 'strain energy' of the cable system in which T_i is the change of tension in a pair of equivalent hangers; $\delta U_D = EI \int_0^l (\mathrm{d}^2\Delta/\mathrm{d}x^2) \, \delta(\mathrm{d}^2\Delta/\mathrm{d}x^2) \, \mathrm{d}x$, the variation of the strain energy of the uniform deck of span l; F_j is a live load; Δ, Δ_i and Δ_j represent deflections of the deck and cable systems at any point, at the position of the ith pair of equivalent hangers and at the jth component of live load, respectively, assuming that the elasticity of the hangers is negligible.

By means of the stiffness coefficients of the cable system relating to the points of connection of the equivalent hangers, the change in tension T_i of the ith pair of hangers due to live load may be expressed in terms of small deflections of the cable system as follows:

$$T_i = b_{i1}\Delta_i + b_{i2}\Delta_2 + b_{i3}\Delta_3 + \cdots + b_{ii}\Delta_i + \cdots + b_{in}\Delta_n \qquad (8.35)$$

so that:

$$\delta U_{\text{C}} = \sum^n (b_{i1}\Delta_1 + b_{i2}\Delta_2 + \cdots + b_{in}\Delta_n)\,\delta\Delta_i \qquad (8.36)$$

Therefore, finally:

$$\delta U = \sum^n (b_{i1}\Delta_1 + b_{i2}\Delta_2 + \cdots + b_{in}\Delta_n)\,\delta\Delta_i$$
$$+ EI \int_0^l \left(\frac{\mathrm{d}^2\Delta}{\mathrm{d}x^2}\right)\delta\left(\frac{\mathrm{d}^2\Delta}{\mathrm{d}x^2}\right)\mathrm{d}x = \sum^N F_j\,\delta\Delta_j \quad (8.37)$$

The procedure of 'exact' solution is to derive n simultaneous equations in the unknown deflections $\Delta_1, \Delta_2, \ldots, \Delta_n$ but a more rapid procedure is to choose a function to represent the deflection of the deck (and cable system) which satisfies the boundary conditions of deflection and which contains only a few unknown parameters. Thus for the main span of a bridge a trigonometrical series is suitable as follows:

$$\Delta = A_1 \sin\frac{\pi}{l}x + A_2 \sin\frac{2\pi}{l}x + A_3 \sin\frac{3\pi}{l}x \qquad (8.38)$$

so that:

$$\Delta_i = A_1 \sin\frac{\pi}{l}x_i + A_2 \sin\frac{2\pi}{l}x_i + A_3 \sin\frac{3\pi}{l}x_i$$

$$\delta\Delta_i = \delta A_1 \sin\frac{\pi}{l}x_i + \delta A_2 \sin\frac{2\pi}{l}x_i + \delta A_3 \sin\frac{3\pi}{l}x_i$$

$$\delta\Delta_j = \delta A_1 \sin\frac{\pi}{l}x_j + \delta A_2 \sin\frac{2\pi}{l}x_j + \delta A_3 \sin\frac{3\pi}{l}x_j \qquad (8.39)$$

$$\frac{\mathrm{d}^2\Delta}{\mathrm{d}x^2} = -\frac{\pi^2}{l^2}\left(A_1 \sin\frac{\pi}{l}x + 4A_2 \sin\frac{2\pi}{l}x + 9A_3 \sin\frac{3\pi}{l}x\right)$$

$$\delta\left(\frac{\mathrm{d}^2\Delta}{\mathrm{d}x^2}\right) = -\frac{\pi^2}{l^2}\left(\delta A_1 \sin\frac{\pi}{l}x + 4\delta A_2 \sin\frac{2\pi}{l}x + 9\delta A_3 \sin\frac{3\pi}{l}x\right)$$

By substituting from equations (8.38) and (8.39) in equation (8.37) and subsequently obtaining the three simultaneous equations of equilibrium generalised for the parameters A_1, A_2 and A_3 by taking the partial derivatives:

$$\frac{\partial U}{\partial A_1} = \sum^N F_j \frac{\partial\Delta_j}{\partial A_1}; \qquad \frac{\partial U}{\partial A_2} = \sum^N F_j \frac{\partial\Delta_j}{\partial A_2}; \qquad \frac{\partial U}{\partial A_3} = \sum^N F_j \frac{\partial\Delta_j}{\partial A_3} \qquad (8.40)$$

6+

the three parameters may be found by simultaneous solution of these equations. Greater accuracy may, of course, be obtained by using additional terms of the series (8.38), the use of more than five terms is, however, rarely justified.

Tables for calculating the stiffness (and flexibility) coefficients for cable systems of suspension bridges are available (bibliography, ref. 23). By using them with the energy procedure described, results within a few per cent of the correct values have been obtained (by M. Svehla and the author) for some major suspension bridges, including that over the Firth of Forth, considering the main span alone. The correct values are those obtained by the non-linear, so-called deflection theory[1] of suspension bridges. The method presented herein provides a lower bound (para. 8.2) for maximum deflection due to live load and an upper bound for the maximum bending moment in the deck[2], the latter being obtained from the results of the deflection analysis. An upper bound for maximum deflection and a lower bound for maximum bending moment may be obtained by the approximate method using complementary energy (para. 8.6). That method is, however, usually too laborious for structures of this kind and accordingly is not recommended.

A less accurate, though more rapid solution of a suspension bridge system may be obtained by neglecting all except the stiffness coefficients $b_{11}, b_{22}, \ldots, b_{ii}, \ldots, b_{nn}$ so that:

$$T_i = b_{ii}\Delta_i \qquad (8.41)$$

for example. The problem is then merely one of a simply supported beam (the deck) with n intermediate discrete elastic supports at the points of attachment of the pairs of equivalent hangers. Its solution follows lines which are identical to those given in para. 8.3 for the system shown in Fig. 8.2.

8.5 The total potential energy method

The procedure of analysis by the total potential energy method (para. 1.4) is identical to that of the conservation of energy method described in paras. 8.2 and 8.3 above. There is, in fact, no conceptual difference between the methods. The total potential energy method is, however, the more formal of the two and provides means of investigating the stability of equilibrium which are outside the scope of this book.

The total potential energy of an elastic (linear or non-linear) structural system, including gravity loading, is:

$$V = U - \sum_{j}^{N} F_j\Delta_j + k \qquad (8.42)$$

---

[1] See bibliography, ref. 22.
[2] These remarks apply to bounds relating to correct solution of the *simplified* structure. As noted above, the effect of simplification of the structure is to offset the error implicit in the conservation and potential energy methods.

where U is the strain energy, k is a constant depending upon the arbitrary choice of the datum of potential energy and the negative term represents the loss of potential of gravity loading during its application to the structure. For a condition of equilibrium the total potential energy of an elastic or con-servative system is stationary (bibliography, ref. 19), that is:

$$\delta V = \delta U - \sum_{}^{N} F_j \, \delta\Delta_j = 0 \qquad (8.43)$$

whence

$$\delta U = \sum_{}^{N} F_j \, \delta\Delta_j \qquad (8.44)$$

as in paras. 8.2 and 8.3.

If, having chosen a suitable function to represent the deflection of a struc-ture which contains, say, independent parameters A_1, A_2 and A_3, then δU and $\delta\Delta_j$ $(j = 1, 2, \ldots, N)$ are functions of these parameters, so that for equili-brium:

$$\frac{\partial V}{\partial A_1} = \frac{\partial U}{\partial A_1} - \sum_{}^{N} F_j \frac{\partial\Delta_j}{\partial A_1} = 0$$

$$\frac{\partial V}{\partial A_2} = \frac{\partial U}{\partial A_2} - \sum_{}^{N} F_j \frac{\partial\Delta_j}{\partial A_2} = 0 \qquad (8.45)$$

$$\frac{\partial V}{\partial A_3} = \frac{\partial U}{\partial A_3} - \sum_{}^{N} F_j \frac{\partial\Delta_j}{\partial A_3} = 0$$

which are identical with equations (8.40) for the conservation of energy method.

It is, therefore, immaterial which of the two methods is used. The conserva-tion of energy method has the possible advantage that its use depends directly upon a well-known law and no formal proof is necessary.

8.6 The complementary energy method for the approximate analysis of statically indeterminate systems

By virtue of the stationary property of complementary energy (paras. 1.4 and 3.8) with respect to the forces in the redundants of a statically-indeterminate system, approximate solutions to a variety of relatively complex problems may be obtained. Such problems include those considered in paras. 8.3 and 8.4 and in order to illustrate the procedure the problem illustrated in Fig. 8.1 and solved by the conservation of energy method in para. 8.3 is convenient.

It is necessary first to describe the distribution of force in the redundants by a function with undetermined parameters which satisfies the boundary con-ditions of equilibrium. This is a general feature of the method. In this parti-cular problem the bending moment M in the beam is representative of the redundants and the boundary condition of equilibrium is that $M = 0$ at $x = l$. The bending moment distribution is, however, discontinuous owing to the intermediate, elastic point supports and for this reason a satisfactory approximation is unlikely to be found easily. Alternatively, the distribution of force in the redundant elastic supports (which is clearly free from discontinuity) may be approximated in a manner which is satisfactory in respect of the boundary conditions of equilibrium. This latter approach is also the more convenient since a function similar to that chosen in para. 8.3 for the deflection of the beam is appropriate, as follows:

$$R_i = R\left(1 - \cos\frac{\pi}{2l}x_i\right) \tag{8.46}$$

where R is a parameter to be determined and x_i is the distance from the origin of the ith elastic support.

Now:

$$\delta C = 0 = \frac{1}{EI}\int_0^l M\,\delta M\,\mathrm{d}x + a\sum^n R_i\,\delta R_i \tag{8.47}$$

where the variation is taken with respect to the redundants or parameters representing them; $EI = 1\cdot5 \times 10^6$ kN m^2; $a = 1/b = 10^{-3}$ m/kN; $l = 24$ m; $n = 6$. In this instance there is a single parameter R representing the redundants, therefore:

$$\frac{\mathrm{d}C}{\mathrm{d}R} = 0 = \frac{1}{EI}\int_0^l M\frac{\mathrm{d}M}{\mathrm{d}R}\,\mathrm{d}x + a\sum^n R_i\frac{\mathrm{d}R_i}{\mathrm{d}R} \tag{8.48}$$

provides the single equation which is necessary for the determination of R. (If the approximating function contained two or more parameters R_p, R_q, etc., the necessary simultaneous equations would be obtained by the partial derivatives $\partial C/\partial R_p = 0$; $\partial C/\partial R_q = 0$, etc.)

For equilibrium the bending moment between each elastic support is as follows, taking account of equation (8.46) and substituting the appropriate values of x_i:

$$M_6 \quad = 0$$
$$M_{5.6} = (R - 10^2)(24 - x)$$
$$M_{4\,5} = (R - 10^2)(24 - x) + 0\cdot741(20 - x)R$$
$$M_{3.4} = (R - 10^2)(24 - x) + 0\cdot741(20 - x)R + 0\cdot5(16 - x)R$$
$$M_{2.3} = (R - 10^2)(24 - x) + 0\cdot741(20 - x)R$$
$$\qquad\qquad + 0\cdot5(16 - x)R + 0\cdot293(12 - x)R \quad (8.49)$$
$$M_{1.2} = (R - 10^2)(24 - x) + 0\cdot741(20 - x)R$$
$$\qquad +0\cdot5(16 - x)R + 0\cdot293(12 - x)R$$
$$\qquad\qquad +0\cdot134(8 - x)R$$
$$M_{0.1} = (R - 10^2)(24 - x) + 0\cdot741(20 - x)\,R$$
$$\qquad +0\cdot5(16 - x)R + 0\cdot293(12 - x)R$$
$$\qquad\qquad +0\cdot134(8 - x)R + 0\cdot034(4 - x)R$$

Thus, with reference to equation (8.48):

$$\int_0^l M \frac{dM}{dR}\, dx = \int_0^{l/6} M_{0-1} \frac{dM_{0-1}}{dR}\, dx + \int_{l/6}^{l/3} M_{1-2} \frac{dM_{1-2}}{dR}\, dx + \cdots$$

$$+ \int_{5l/6}^l M_{5-6} \frac{dM_{5-6}}{dR}\, dx \quad (8.50)$$

and

$$\frac{dR_i}{dR} = \left(1 - \cos \frac{\pi}{2l} x_i\right) \tag{8.51}$$

where $l = 24$ m.

Substituting from equations (8.49), (8.50) and (8.51) in equation (8.48) provides an equation from which is obtained $R = R_6 = 44 \cdot 1$ kN which should be compared with the correct value of $46 \cdot 5$ kN. This result indicates the lower bound to R_6 which is to be expected from the complementary energy method because approximation of the statically indeterminate force distribution implies failure to satisfy compatibility conditions precisely, i.e. an increase in flexibility of the system. In this instance, however, the lower bound on maximum force corresponds with a lower bound on the maximum deflection due to the nature of the approach to the problem. It is, therefore important to exercise caution in respect of conclusions relating to upper and lower bounds, especially because, in general, the complementary energy approach is ill-conditioned.

The complementary energy procedure is undoubtedly lengthier and more tedious for problems of this kind than its conservation of energy or total potential energy, counterpart. Also, since it is not so easy to visualise distribution of force or bending moment as it is to visualise deflection for a given system, the chances of making a good choice of approximating function for use in the complementary energy procedure are relatively poor (except, of course when the approximate solution by the alternative procedure is available). Unless, therefore, it is necessary to obtain both upper and lower bounds to the solution of a problem the latter methods will generally prove to be the more convenient. Application of the approximate complementary energy method to the suspension bridge problem is treated by C. F. P. Bowen and the author elsewhere (see bibliography, ref. 23).

EXERCISES

1 Using an approximate method estimate the deflection at the centre of a uniform beam of flexural rigidity EI and span l, encastré at both ends, due to a uniformly distributed load of intensity w per unit length of span. Estimate also the restraining couples at the ends of the beam due to that loading.

Ans: $wl^4/384EI$ } Correct
$\pm\ wl^2/12$ } values

2 A uniform beam of flexural rigidity $EI = 10^7$ kN cm^2 and length 6 m is supported at each end and at five equally spaced intermediate points by elastic supports each of stiffness 1 kN/cm. Estimate the deflection at one end of the beam due to a force of 10 kN applied there. (Try, for example, $\Delta = A_1 + A_2 x + A_3 x^3$.)

Ans: 5·2 cm

3 Analyse the system shown in Fig. 8.1 approximately by the complementary energy method assuming that the distribution of bending moment is given by $M = A_1 - A_2 x + A_3 \sin(\pi/l)x$. (Note: one of the parameters may be eliminated by applying the boundary condition of equilibrium.)

Ans: $M_{max} = 290$ kN m

9

Aspects of computation

9.1 For various reasons there was an energetic movement away from formal methods of framework analysis in this country and the U.S.A. following the publication of Professor Hardy Cross's paper [1] on moment distribution (based implicitly on the equilibrium approach) in 1930. This movement received additional impetus and embraced a wider field with Southwell's introduction of relaxation methods in 1935. [2] In recent years the coming of automatic digital computing machines and the desire of engineers to make use of their enormous potentialities in the interests of economy, has tended to reverse this trend. This is due to the fact that digital computers are well suited to performing the processes of formal analysis and on the basis of the equilibrium approach, a series of instructions or 'program' of universal utility for framework analysis, is possible for these machines. As well as eliminating manual labour, the digital computer has, therefore, led to a long-overdue reappraisal of framework analysis, particularly in respect of the value of working in terms of stiffness and flexibility coefficients for systems with linear elasticity. It is interesting that formal procedures retained their popularity on the European Continent.

9.2 Appreciation of formal analysis

Formal analysis of structures whereby the final equations representing either conditions of equilibrium or compatibility of strains are set up and solved systematically, has advantages in addition to those in respect of computer programming. In the event of more than one condition of loading of a framework being involved (which is usual in practice) the formal procedure is easily modified to take account of each of the conditions. Thus, it is merely necessary to recalculate the load column only of the tabular solution of the simultaneous equations for each condition of loading (see para. 6.6). Moreover, formal analysis cultivates thorough understanding of the analytical processes and, in consequence, a progressive, critical approach to the treatment of

[1] See bibliography, ref. 6.
[2] See bibliography, ref. 5.

different kinds of framework. The undesirable feature of formal analysis for manual purposes is the necessity of solving a set of simultaneous equations. This process can, however, be reduced to a routine procedure by using the method of successive elimination in tabular form as shown in para. 6.6 and the operation is performed only once, regardless of the number of loading conditions. Adoption of an iterative procedure such as relaxation, is at the expense of this universal feature and furthermore, the true power of relaxation cannot be exploited in this way.

9.3 Use of matrices in structural analysis

The mathematical device of matrices[1] has been found to be convenient for describing and performing the processes of formal analysis of complicated structures whose elasticity is linear. This is true, particularly in relation to the programming of digital computers for structural analysis. Matrix methods are outside the scope of this book but some of their essential features can be demonstrated in simple terms.

The final equations of the equilibrium approach to the analysis of a framework of any kind have the following form:

$$
\begin{aligned}
F_1 &= b_{11}\Delta_1 + b_{12}\Delta_2 + \cdots + b_{1n}\Delta_n \\
F_2 &= b_{21}\Delta_1 + b_{22}\Delta_2 + \cdots + b_{2n}\Delta_n \\
&\;\vdots \\
F_n &= b_{n1}\Delta_1 + b_{n2}\Delta_2 + \cdots + b_{nn}\Delta_n
\end{aligned}
\tag{9.1}
$$

where the loads F are expressed in terms of the components of deflection Δ of joints (unknown) and the relevant stiffness coefficients b, n being the number of degrees of freedom of deflection of joints.

These equations may be expressed in terms of matrices as follows:

$$
\begin{bmatrix}
b_{11} & b_{12} & \cdots & b_{1n} \\
b_{21} & b_{22} & \cdots & b_{2n} \\
\vdots & \vdots & & \vdots \\
b_{n1} & b_{n2} & \cdots & b_{nn}
\end{bmatrix}
\begin{bmatrix}
\Delta_1 \\
\Delta_2 \\
\vdots \\
\Delta_n
\end{bmatrix}
=
\begin{bmatrix}
F_1 \\
F_2 \\
\vdots \\
F_n
\end{bmatrix}
\tag{9.2}
$$

This is, the product of the stiffness (square) matrix, a scalar quantity, and the deflection vector (column matrix) is equal to the force or load vector. The arrangement of the matrices in equation (9.2) is important since a square matrix must be post-multiplied by a column matrix or vector. The way in which this operation is carried out is apparent when equations (9.1) and (9.2) are compared. Equations (9.2) may be expressed in abbreviated notation, thus:

$$b.\Delta = F$$

or

$$\Delta = b^{-1}F = aF$$

(9.3)

[1] Matrices were used by Müller-Breslau.

where a, the inverse of the stiffness matrix, is a flexibility matrix with the same number of rows (and columns):

$$\begin{bmatrix} b_{11} & b_{12} & \cdots & b_{1n} \\ b_{21} & b_{22} & \cdots & b_{2n} \\ \vdots & \vdots & & \vdots \\ b_{n1} & b_{n2} & \cdots & b_{nn} \end{bmatrix}^{-1} = \begin{bmatrix} a_{11} & a_{12} & \cdots & a_{1n} \\ a_{21} & a_{22} & \cdots & a_{2n} \\ \vdots & \vdots & & \vdots \\ a_{n1} & a_{n2} & \cdots & a_{nn} \end{bmatrix} \qquad (9.4)$$

Thus, equations (9.1) may be solved by inverting the stiffness matrix, a process which is described in textbooks on matrix methods.[1] It is important to note that while the form of equations (9.3) is similar to that of a system having one degree of freedom the individual elements of the stiffness and flexibility matrices do not bear a simple relationship one to another, for example $1/b_{ij} \neq a_{ij}$.

By the compatibility approach, however, the final equations are generally as follows:

$$\begin{aligned} a_{11}T_1 + a_{12}T_2 + \cdots + a_{1r}T_r &= \Delta_1 \\ a_{21}T_1 + a_{22}T_2 + \cdots + a_{2r}T_r &= \Delta_2 \\ &\vdots \\ a_{r1}T_1 + a_{r2}T_2 + \cdots + a_{rr}T_r &= \Delta_r \end{aligned} \qquad (9.5)$$

where the coefficients a of the forces or couples T of the r redundants are flexibility coefficients of the chosen statically determinate system (but for a truss $a_{11}, a_{22}, \ldots, a_{rr}$ include the flexibility of the respective redundant members) and $\Delta_1, \Delta_2, \ldots, \Delta_r$ represent the negatives of the deformations which the specified system of loads would cause in the absence of redundants, at the locations of the chosen set of redundants. Thus, in terms of matrices:

$$\begin{bmatrix} a_{11} & a_{12} & \cdots & a_{1r} \\ a_{21} & a_{22} & \cdots & a_{2r} \\ \vdots & \vdots & & \vdots \\ a_{r1} & a_{r2} & \cdots & a_{rr} \end{bmatrix} \begin{bmatrix} T_1 \\ T_2 \\ \vdots \\ T_r \end{bmatrix} = \begin{bmatrix} \Delta_1 \\ \Delta_2 \\ \vdots \\ \Delta_r \end{bmatrix} \qquad (9.6)$$

that is:

$$aT = \Delta$$

or

$$T = a^{-1}\Delta \qquad (9.7)$$

The redundant forces or couples may, therefore, be found by inverting the flexibility matrix and post-multiplying by the deformation vector. It should be noted that the flexibility matrix and deflection vector bear no relationship to the stiffness matrix and deflection vector which relate to the analysis of the same framework by the equilibrium approach. These latter refer to the stiffness and deflections of joints, respectively, of the statically indeterminate systems.

[1] See bibliography, ref. 16.

6* +

Matrices are simply groups of similar quantities for which simple arithmetical rules have been devised to enable such groups to be manipulated as though they were single quantities or variables. They enable all except the final stage of the analysis of complex linear systems to be dealt with in the simplest terms.

9.4 Use of digital computers for framework analysis

A digital computing machine can be instructed or 'programmed' to carry out automatically the whole process of framework analysis. Moreover, the set of coded instructions or the 'program' can be of universal utility for frameworks. For each framework, whether it is a pin-jointed truss or a complicated rigidly jointed portal framework, it is merely necessary to preface the program with the data of the framework and its loading.

The design of a universal program for framework analysis involves consideration of two important factors. One is, perhaps, obvious; that is that the method of analysis upon which the program is based must be perfectly general, regardless of the nature of the framework. The other is that it must be capable of description as a sequence of operations which can be followed blindly, regardless of the nature of the framework, as a simple routine. Both of these considerations are admirably satisfied by the equilibrium approach which, unlike the compatibility approach, involves no arbitrary choice of unknown quantities during the course of the analysis. It will be recalled that an arbitrary choice in respect of redundants is necessary during the course of the compatibility approach. Thus, in spite of the compatibility approach being the more convenient for manual analysis in some instances, the considerations which determine this are irrelevant to computer programming. For example, the digital computer is insensitive to the number of simultaneous equations involved.

Adopting the equilibrium approach, the computer program makes use of coded data supplied to the machine to perform the following operations separately:

(a) formation of the stiffness coefficients of the individual members;
(b) formation of the stiffness coefficients of the framework as a whole;
(c) formation of the final simultaneous equations of equilibrium;
(d) solution of the simultaneous equations;
(e) printing-out the results (deflections) of the analysis or their conversion to bending moments at the ends of members followed by printing out of both deflections and bending moments.

These operations are achieved by instructions contained in separate sections of the program called 'routines' or 'sub-routines'. Several sub-routines of a framework analysis program are common to other programs concerned with linear analysis. When they are not in use, the sub-routines are kept in a 'program library' where they are available for the making-up of programs generally.

It is not necessary for the engineer to be able to program digital computers because these machines are operated by a staff of specialists. In order that he can provide the data for the solution of his particular problem in a suitable form, it is, however, desirable that the engineer understands the mathematical basis of the relevant program.

9.5 The relaxation method for linear systems

The essential features of the mathematical process of relaxation may be demonstrated by using it to solve linear simultaneous equations. For example, in order to solve the equations:

$$F_1 = 10 = 4\Delta_1 + 2\Delta_2$$
$$F_2 = 20 = 2\Delta_1 + 5\Delta_2 \tag{9.8}$$

they are rewritten as follows:

$$R_1 = 10 - 4\Delta_1 - 2\Delta_2$$
$$R_2 = 20 - 2\Delta_1 + 5\Delta_2 \tag{9.9}$$

where $R_1 = R_2 = 0$ for the correct values of Δ_1 and Δ_2 that is, the values which satisfy equations (9.8). For any other values of Δ_1 and Δ_2 the magnitudes of R_1 and R_2, called 'residuals', are indicative of the deviation from the correct solution. Relaxation consists of successively correcting Δ_1 and Δ_2 until the residuals become insignificant for practical purposes. For the purpose it is desirable to construct an operations table (Table 9.1) to indicate the effect on the residuals of unit changes of each of the variables in turn. Thus, for the equations (9.8) by referring to operation table 9.1 the process of relaxation can be carried out in tabular form, as in Table 9.2.

TABLE 9.1

	δR_1	δR_2
$\delta\Delta_1 = 1$	-4	-2
$\delta\Delta_2 = 1$	-2	-5

Changes in Δ_1 and Δ_2 are considered successively which reduce R_1 and R_2 to zero successively. In this example, a change in Δ_2 is considered first because R_2 is the greater residual. As the process proceeds, the alternate values of R_1

and R_2 become smaller and when they are insignificant relaxation is discontinued. The solution of the equations to the desired accuracy is then obtained by adding the $\delta\Delta_1$s and $\delta\Delta_2$s in the first and second columns of Table 9.2 as shown. In this instance relaxation has been discontinued when $R_1 = 0$ and $R_2 = -0.04$.

TABLE 9.2

Operation	R_1	R_2
$\Delta_1 = \Delta_2 = 0$	10	20
$\delta\Delta_2 = 4$	2	0
$\delta\Delta_1 = 0.5$	0	-1
$\delta\Delta_2 = -0.2$	0.4	0
$\delta\Delta_1 = 0.1$	0	-0.2
$\delta\Delta_2 = -0.04$	0.08	0
$\delta\Delta_1 = 0.02$	0	-0.04
$\Delta_1 = 0.62; \Delta_2 = 3.76$	0	-0.04

It is not necessary to adhere to the precise procedure and variations can be practised which seem to accelerate convergence. Thus, simultaneous changes in Δ_1 and Δ_2 may be made, or again, changes which cause reversal of sign of one or more residuals at particular stages. In fact, relaxation is a process which can be adapted readily to make use of the ability and foresight of the individual and its rapidity for obtaining solutions depends largely upon the computor. It was never intended by its originator, Sir Richard Southwell, F.R.S. to be merely an iterative method of solving linear simultaneous equations. He described it in terms of a physical concept, the systematic relaxation of constraints. Thus, if equations (9.8) are the final equations of equilibrium of the framework shown in Fig. 3.1, R_1 and R_2 actually represent external 'constraints' which must be in action to maintain equilibrium if Δ_1 and Δ_2 do not correspond with the loading F_1 and F_2. In other words, R_1 and R_2 are external influences which must be present at any stage to assist the framework in supporting the loads. The objective of relaxation is the systematic liquidation of such constraints and it can be done without ever setting up the relevant simultaneous equations formally.

It is interesting to note that any numerical errors which are made during relaxation merely affect the rate of convergence and do not cause the final result to be erroneous. In other words, errors introduce additional constraint which is eventually relaxed. Unfortunately, the relaxation procedure must be repeated entirely for each condition of loading of a framework and the numerical effort required becomes enormous for complicated systems. The method of moment distribution for rigidly jointed frameworks is a form of relaxation method where attention is concentrated upon the bending moments at the ends of members during the process instead of on the residual couples.

It suffers from the same disadvantages as the usual form of relaxation plus the additional one that errors made during the process persist and lead to incorrect results, due to ignorance of the residuals throughout.

9.6 Ill-conditioned simultaneous equations

The solution of simultaneous equations is extremely tedious if they are 'ill-conditioned'. For example, the linear simultaneous equations:

$$5\Delta_1 + 7\Delta_2 + 6\Delta_3 + 5\Delta_4 = 23 = F_1$$
$$7\Delta_1 + 10\Delta_2 + 8\Delta_3 + 7\Delta_4 = 32 = F_2$$
$$6\Delta_1 + 8\Delta_2 + 10\Delta_3 + 9\Delta_4 = 33 = F_3$$
$$5\Delta_1 + 7\Delta_2 + 9\Delta_3 + 10\Delta_4 = 31 = F_4$$

(9.10)

are satisfied exactly by $\Delta_1 = \Delta_2 = \Delta_3 = \Delta_4 = 1$. It may be shown, however, that the following values satisfy the equations to the accuracies shown:

TABLE 9.3

Δ_1	Δ_2	Δ_3	Δ_4	F_1	F_2	F_3	F_4
14·6	−7·2	−2·5	3·1	23·1	31·9	32·9	31·1
2·36	0·18	0·65	1·21	23·01	31·99	32·99	31·01
1·136	0·918	0·965	1·021	23·001	31·999	32·999	31·001

Thus, unless the equations are solved to a high degree of accuracy, the solution is likely to be grossly in error. Equations of this kind are said to be 'ill-conditioned'. The example shown is due to T. S. Wilson.[1]

Equations are liable to be ill-conditioned if the coefficients on the leading diagonal (the self-coefficients $b_{11} = 5$, $b_{22} = 10$, $b_{33} = 10$ and $b_{44} = 10$ in the example considered) are of similar magnitude to the other coefficients (the mutual coefficients). Conversely, linear simultaneous equations are usually well-conditioned if the self-coefficients are large in comparison with the mutual coefficients.

The physical significance of ill-conditioning can be appreciated by referring again to equations (9.10) for which it is clear from Table 9.3 that small changes in the Fs cause large changes in the Δs. Thus, if these equations represent the final equations of equilibrium of a framework, it would be such that small changes in its loading would cause large changes in its deflections. A framework of this kind would be useless for most practical purposes. It is, therefore unusual for the equilibrium approach to lead to ill-conditioned equations. If,

[1] See bibliography, ref. 21.

on the other hand, the equations of the compatibility approach are ill-conditioned this signifies that small changes in deflections require large changes in the forces to produce them. This state of affairs is common in frameworks of high rigidity and is a desirable feature. The equations of the compatibility approach are, therefore, quite likely to be ill-conditioned.

Solution of ill-conditioned equations by the relaxation method is likely to be well nigh impossible due to the poor rate of converge in such instances. This can, however, be remedied by means of the 'multiplying factor' method, details of which can be found elsewhere (see, for example, bibliography, ref. 15).

Appendix I

History of structural analysis[1]

Navier was responsible for the first treatment of a statically indeterminate framework. Fig. 1 shows the plane, pin-jointed framework which formed Fig. 112 of his remarkable book published in 1826.[2] He stated: 'When a load is supported by more than two inclined members in the same vertical plane or by more than three inclined members not in the same plane, the conditions of equilibrium leave undetermined between certain limits, the forces impressed in the direction of each of the members.' Subsequently he noted that since the members of the structure are capable of deformation in a way dictated by the elasticity of their material the distribution of load between the members is no longer indeterminate. This he demonstrated by supposing that p, p' and p'' are the forces exerted by the members in resisting the weight II, whence for equilibrium:

$$p \cos a + p' \cos a' + p'' \cos a'' = \text{II}$$
$$-p \sin a + p' \sin a' + p'' \sin a'' = 0$$

(1)

Now, according to Navier, the changes in length of the members AC, A'C and A''C, consistent with their being connected at C, are $f \cos a + h \sin a$, $f \cos a' - h \sin a'$ and $f \cos a'' - h \sin a''$, where f and h are the (small) vertical and horizontal components of the deflection of C, so that the compressive strains of the members are:

$$\frac{f \cos^2 a + h \sin a \cos a}{a}$$

$$\frac{f \cos^2 a' - h \sin a' \cos a'}{a}$$

and

$$\frac{f \cos^2 a'' - h \sin a'' \cos a''}{a}$$

respectively, a being the height of C above AA''.

[1] Based on an article by the author in *Civil Engineering and Public Works Review*, **61**, 1966, 203–7, by kind permission of the publishers of that journal.
[2] See ref. 1.

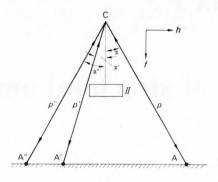

Fig. 1

Hence:

$$p = F \cdot \frac{f \cos^2 \alpha + h \sin \alpha \cos \alpha}{a}$$

$$p' = F' \cdot \frac{f \cos^2 \alpha' - h \sin \alpha' \cos \alpha'}{a} \tag{2}$$

$$p'' = F'' \cdot \frac{f \cos^2 \alpha'' - h \sin \alpha'' \cos \alpha''}{a}$$

where F, F' and F'' are the elastic constants of the members respectively, which he said, 'together with the two equations 1, will give the values of the displacements h and f and the forces p, p', p''.' It is noteworthy that Navier's exposition lacks clarity in respect of the positive senses of h and f and that, in any case, there were errors of sign in his equations (thus $p \sin \alpha$ was a positive term in the second of his equations of equilibrium and he wrote the change in the length of AC as $f \cos \alpha - h \sin \alpha$ though it seems likely that he intended f to be positive downward and h positive to the right). Nevertheless, he set down the essential features of the most general approach to the analysis of statically indeterminate frameworks. Moreover, Navier's book contained a correct treatment of the analysis of continuous beam systems. A substantial part of Navier's work was made available to the English-speaking world by the Rev. H. Moseley (Professor of Natural Philosophy at King's College, London) in his book *The Mechanical Principles of Engineering and Architecture* of 1843.

The German mathematician Clebsch, Professor at the Polytechnic Institute, Karlsruhe, adopted the same general approach to the analysis of pin-jointed frameworks. This he published in a book on the theory of elasticity which appeared in 1862 and which was later translated into French by St. Venant and Flamant with annotations (1883),[1] the former claiming in a footnote (page 871) to have given the general principle of this method of structural analysis in his course notes of 1838 at l'École des Ponts et Chaussées. Clebsch

[1] See ref. 2.

presented the equations of equilibrium of the single 'free' joint of a linearly elastic space framework of the kind shown in Fig. 2 in the following form:

$$X = b_{xx}u + b_{xy}v + b_{xz}w$$
$$Y = b_{yx}u + b_{yy}v + b_{yz}w \tag{3}$$
$$Z = b_{zx}u + b_{zy}v + b_{zz}w$$

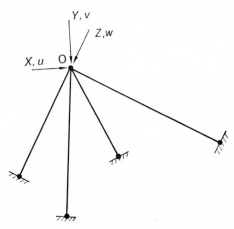

Fig. 2

In these linear equations u, v and w are the components of the (small) deflection of the free joint caused by the loads, X, Y and Z. In order to obtain these equations he first expressed the forces in the members in terms of u, v and w using the strain compatibility conditions and the elastic constants of the members. Then he substituted the expression so obtained in the three equations of equilibrium of the free joint and finally combined the coefficients of u, v and w, respectively, in each equation and denoted them by b_{xx}, etc. (actually he used the symbol a for these combined coefficients) to obtain the form of equations (3). Now these coefficients are the stiffness coefficients of the structure and, as Clebsch noted, $b_{xy} = b_{yx}$, etc., a property which was later to be identified with the reciprocal theorem for linear systems. If he had stated the physical significance of this property he might well have been credited with the discovery of the reciprocal theorem. As it happened, Maxwell's interpretation of this feature of the behaviour of linear structures in his celebrated paper of 1864 earned him that distinction.

It is remarkable that except for its application to complicated frameworks, including those with rigid joints, Clebsch's treatment of the problem is identical with that which has been found the most suitable for programming automatic computers for structural analysis. This general approach (in common with the alternative) is, moreover, inherently unrestricted and can be used for structures made of material which does not obey Hooke's law, that is, non-linear, but whose deflections are small. Then, however, it is not possible to introduce the usual concept of stiffness coefficients, the final equations of equilibrium in the displacements being non-linear. An important feature is that there are as many final equations (of equilibrium with components of deflection as unknowns) as there are degrees of freedom of deflection of the structure.

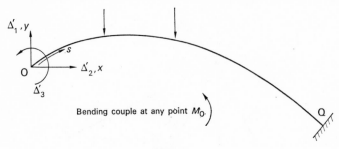

Fig. 3

An important development had, however, taken place before the appearance of Clebsch's book. This was the publication in Paris in 1854 of Bresse's book *Researches analytiques sur la flexion et la resistance des pieces combes*. The deflection of elastic arches was one of the topics treated by Bresse and he derived the equations for the components of deflection of the free end of an arch encastré in one abutment. When deformation due to bending only is taken into account, Bresse's equations assume the well-known forms (Fig. 3):

$$\Delta'_1 = -\int_0^Q \frac{M_0 x \, ds}{EI}$$

$$\Delta'_2 = \int_0^Q \frac{M_0 y \, ds}{EI} \tag{4}$$

$$\Delta'_3 = \int_0^Q \frac{M_0 \, ds}{EI}$$

He noted, moreover, that the statically indeterminate arch encastré in two abutments could be analysed by using the condition for $\Delta_1 = \Delta_2 = \Delta_3 = 0$ at an abutment. Here we see for the first time the approach to the analysis of a statically indeterminate structure whereby the final equations represent conditions of compatibility of strain, having taken cognisance of the conditions of equilibrium in their formulation with redundant forces and couples as unknowns. This approach has been called by Charlton the 'compatibility approach' in contrast with that used by Navier and Clebsch which he has called the 'equilibrium approach' (1956).[1] (It is noteworthy, incidentally, that Navier used the compatibility approach in his book of 1826 for the

[1] See ref. 26. This terminology is preferred because *both* approaches can have forces or deflections as variables, see ref. 27.

analysis of continuous beam systems.) While the latter is the more generally useful for structures consisting of straight members, the former reigns supreme for arches. Thus, the three redundants (Fig. 4) of an encastré arch, say R_1, R_2 and R_3 may be determined from the equations

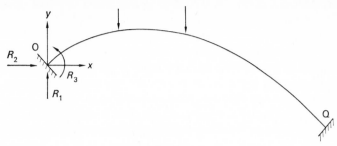

Fig. 4

$$\Delta_1 = \Delta_1' + a_{11}R_1 + a_{12}R_2 + a_{13}R_3 = 0$$
$$\Delta_2 = \Delta_2' + a_{21}R_1 + a_{22}R_2 + a_{23}R_3 = 0 \tag{5}$$
$$\Delta_3 = \Delta_3' + a_{31}R_1 + a_{32}R_2 + a_{33}R_3 = 0$$

where:

$$a_{11} = \int_0^Q \frac{x^2 \, ds}{EI}; \quad a_{12} = a_{21} = -\int_0^Q \frac{xy \, ds}{EI}; \quad a_{13} = a_{31} = -\int_0^Q \frac{x \, ds}{EI}$$

$$a_{22} = \int_0^Q \frac{y^2 \, ds}{EI}; \quad a_{23} = a_{32} = \int_0^Q \frac{y \, ds}{EI}; \quad a_{33} = \int_0^Q \frac{ds}{EI}$$

The as are now 'flexibility coefficients' of the linearly elastic arch and the Δ's are the deflections of O which would occur due to the loading if $R_1 = R_2 = R_3 = 0$. The compatibility equations (5) represent, then, the conditions for no deflection of O. It should be noted that they are as numerous as the redundants, a characteristic feature of this approach to the analysis of statically indeterminate systems generally. (Later Culmann proposed the concept of the 'elastic centre' whereby these equations of arch behaviour may be simplified and from which Hardy Cross derived the so-called 'Column Analogy', topics which are treated in Chapter 5.)

It may seem in retrospect that by 1854 the stage had apparently been set for rapid advances in structural analysis for in principle there was nothing new to be discovered. Unfortunately, however, Navier's lead seems to have been overlooked (as was that given later by Clebsch) and Bresse apparently did not enlarge his field of interest to include braced frameworks. Consequently, the solution of statically indeterminate frameworks was still regarded as a challenge when Maxwell (Jenkin having apparently brought the problem to his attention[1]) addressed himself to the problem and published his findings in 1864 in the *Philosophical Magazine*. Maxwell derived a form of the compatibility approach for the purpose of analysing statically indeter-

[1] See ref. 8.

171

minate pin-jointed frameworks. He confined his attention to such frameworks having linear elasticity and behaviour (i.e. small deflections in accordance with the requirements of engineering practice) in common with his contemporaries, and as well as demonstrating a process of analysis he drew attention to the reciprocal property of their deflections. Being unaware of Maxwell's work, Otto Mohr, Professor at Stuttgart, published in 1874 a similar method of analysing statically indeterminate frameworks.[1] His approach differed from Maxwell's only in respect of the formulation of deformations or strains in the process of determining the conditions of compatibility of the strains of the redundants with those of the remainder (statically determinate part) of the structure. Maxwell used a procedure based upon what he called Clapeyron's theorem but which is merely an application of the law of conservation of energy, while Mohr used the more powerful method of the principle of virtual work. (It should be noted, in passing, that the necessary strain compatibility conditions for structures can be obtained geometrically without recourse to energy devices as indeed was shown in principle by Navier.)

The simple framework considered by Navier with its two degrees of freedom and one redundant may be used with advantage to compare his approach, the equilibrium approach, with that of Maxwell and Mohr, the compatibility approach. The compatibility approach is characterised by the fact that a choice in respect of redundants must be made at the outset.

Unfortunately, Maxwell's paper on framework analysis passed unnoticed by the engineering profession[2] in Britain, such was the prevailing attitude to engineering science in these islands at that time and as revealed forcibly by J. B. Chalmers in his book on graphical statics of 1881:

> There are, no doubt, amongst us, a large number, who in earlier years have studied their Pratt, their Navier, their Moseley, or who in more recent years have become familiar with their Bresse and Rankine, have made themselves acquainted with Clapeyron's Theorem of the Three Moments, even a few to whom Lamé is not unknown, but those have done so without hope of reward, simply that they might be truthful men, knowing that which they as Engineers profess to know. But by how many are such surrounded, often jostled, undistinguishable from them by the laity, committing blunders by rule of thumb, affecting to despise science, talking vaguely of their experience and of the practical, whence our public structures suffer in strength, elegance and economy from vicious design, and our public works from defective method in their conception.[3]

[1] In the same year Levy (pupil of St. Venant) had his book (ref. 10) published which included his own treatment of the analysis of statically indeterminate frameworks after Navier, by the compatibility approach as well as studies of weight-saving in frameworks.
[2] Use was, however, made of Maxwell's basic method by Professor Fleeming Jenkin in 1869 (ref. 8). It is interesting to note that unlike Maxwell, Jenkin made use of the principle of virtual work to calculate the relevant structural deformations.
[3] Chalmers, F. B. *Graphical Determination of Forces in Engineering Structures*. Macmillan, London, 1881, pp. xiv–xv.

Mohr's work was eagerly assimilated in Germany, however, where the scientific approach to engineering was well established in common with other countries of the European continent. Nevertheless, it was Castigliano who (following the work of Menabrea) achieved the widest acclaim for his so-called principle of least work for analysing statically indeterminate frameworks. Presented first in a thesis to the University of Turin in 1873, Castigliano's principle of least work was published in 1875 and was acclaimed in Germany by Müller-Breslau (who, incidentally, seems to have priority in introducing the concept of flexibility coefficients in using the Maxwell–Mohr method for linear structures). It achieved the same result for linearly elastic structures as the Maxwell–Mohr method in providing the final equations of the compatibility approach to analysis, but, unlike that method, did so in a manner such that no physical appreciation of the process was necessary on the part of the user. Attention was concentrated on the strain energy of the structure expressed in terms of the loads and forces in the redundant members and its derivatives with respect to the latter. The only obstacle remaining after Castigliano seemed to be the labour of solving the inevitable simultaneous equations.

The appearance of Castigliano's principle was, however, a mixed blessing, for not only did it tend to discourage understanding of structural behaviour but it attracted attention away from the more important general concepts, including virtual work, and introduced the mystique of 'economy of nature' into structural theory. In spite of the corrective influence of Crotti[1] and Engesser in introducing in 1888 and 1889, respectively, the complementary energy concept which embraces Castigliano's principle and is valid for structures made of material whose elasticity is non-linear, it is only comparatively recently that a correct appreciation of energy principles has been at all widely achieved.

The so-called principle of least work was, incidentally, discovered by the English mathematician, J. H. Cotterill, F.R.S., some ten years earlier (1865) than Castigliano. His misfortune was that of publishing his work in England where it was overlooked by those to whom it could have been useful. Cotterill's treatment could, in fact, be held to be preferable on fundamental grounds to that of Castigliano. It is noteworthy that his interest in the subject stemmed originally from Moseley's principle of least resistance relating to the mechanics of the masonry arch (1833).

Following the widespread adoption of Castigliano's method which, incidentally, did not come to pass in Britain until the end of the last century, there appears to have been a succession of piecemeal efforts directed at the analysis of particular types of structures. Thus, secondary stresses in rigidly jointed trusses received a considerable amount of attention between 1880 and 1892, one result of which was the slope-deflection procedure in which Mohr, Manderla and Winkler played a part. Then in 1914 Bendixen published his

---

Based like that of Cotterill on mathematical operations in contrast with Engesser's virtual work approach.

so-called slope-deflection method for analysing rigidly jointed frameworks of the portal type. This utilises the equilibrium approach since the end product is the set of final equations of equilibrium in terms of the deflections as unknowns. It was not until the appearance of Ostenfeld's paper in 1921 (followed by his book in 1926) that a unified outlook on structural analysis emerged, but it seems to have attracted relatively little attention.

With the ever increasing complexity of structures in the twentieth century, including aircraft structures, much effort was devoted to means of removing the drudgery of solving the simultaneous equations of structural analysis. As early as 1868, Winkler and Mohr had introduced the concept of influence lines while in 1884[1] Müller-Breslau had used the reciprocal theorem, it seems, to derive his principle for determining influence lines for the forces in members of linear structures. Contemplation of these contributions led Professor G. E. Beggs in 1922 to propose the use of scale models for obtaining the influence lines for statically indeterminate quantities of plane structures (Chapter 7). Thus, the useful device known as model analysis came into being and has, incidentally, developed into a valuable modern method.

Further relief to the 'stressman' came in 1930 when Professor Hardy Cross introduced his method of moment distribution (which utilises the equilibrium approach to analysis though attention is concentrated on terminal bending moments during the iterative process involved). The principle underlying the Cross method was later generalised by Southwell in the form of his relaxation technique (Chapter 9) which has been adapted to a wide variety of engineering problems. Iterative methods were not unknown before Cross and Southwell but moment distribution and relaxation had the strong appeal of a physical concept or process. It is interesting to note that while these informal methods of computation gained widespread popularity in the English-speaking world, the formal solution of the simultaneous equations of structural analysis was still practised widely on the continent of Europe. This is hardly surprising: the Gauss procedure (Chapter 6) is easy to operate and, moreover, different conditions of loading of a structure can be considered by repeating only a small proportion of the arithmetical work.

Finally, the appearance of automatic digital computers a few years after the second world war led to a long-overdue reappraisal of structural analysis. It soon became apparent that if advantage was to be taken of the enormous capacity of these machines for carrying out complex arithmetical processes for structural design calculations, the somewhat ad hoc methods developed of necessity during the previous fifty years would have to be abandoned. An analytical process was needed which was completely general regardless of the type of structure and which did not vary in its mode of operation. The answer lay in the process suggested by Navier (1826) as later refined by Clebsch and Ostenfeld, the equilibrium approach, whereby final equations of structural analysis are formulated with deflections of joints as unknowns.

[1] *Z. d. Arch. u. Ing. Ver. Z. Hannover* 1884, p. 278.

For linear structures the coefficients of the deflections (rotational and translational) are the relevant stiffness coefficients. The calculation of stiffness coefficients follows a set pattern and the deflections are associated with the degrees of freedom of a structure, so that once the geometry of a structure is defined the other steps follow automatically as indeed does the solution of the final equations. Matrix notation was found to be a convenient vehicle for computer programming, but in essence the computer operation is in complete accord with formal or classical analysis. One of the earliest publications in this field was R. K. Livesley's article of 1953.[1]

Thus, in principle the subject of analysis of statically indeterminate structures has returned to its starting point for the purpose of automation. This has had a beneficial effect in promoting greater depth of understanding of the subject apart from the purely utilitarian aspect. Already in the immediate post-war years there was a vital interest in Britain, albeit on the part of a few, in a re-examination of structural analysis and to appreciate its development in Germany. (There is no doubt that the continentals were better prepared for the computer age, engineer for engineer, than were we in the British Isles.) The result of this activity has been time-saving devices such as those derived from the principle of virtual work and awareness of simplifications afforded by consideration of symmetry and anti-symmetry, a consequence of the principle of superposition for linear systems. For in spite of automatic computers there are still many circumstances in which manual effort is required.

REFERENCES

1 NAVIER, L. M. H. *Resumé des Leçons données à l'Ecole des Ponts et Chaussées sur l'Application de la Mecanique à l'Etablissement des Constructions et des Machines*, Carilian-Goeury, Paris 1826

2 CLEBSCH, A. *Theorie der Elasticität Fester Korper*, 2nd ed. (with notes by Saint-Venant and Flamant), Dunod, Paris 1883

3 MOSELEY, H. *The Mechanical Principles of Engineering and Architecture*, Longmans, London 1843 (contains an attempt to use strain energy derivatives)

4 LAMÉ, G. *Leçons sur la Theorie Mathematique de l'Elasticité des Corps Solides*, Gauthier-Villars, Paris 1852 (Clapeyron is credited by Lamé, p. 91, with using virtual work for finite elastic displacements)

5 MAXWELL, J. C. 'On the calculation of the equilibrium and stiffness of frames', *Phil. Mag.*, 4th ser., **27**, 1864, 294

6 COTTERILL, J. H. 'On an extension of the dynamical principle of least action', *Phil Mag.*, 4th ser., **29**, 1865, 299

7 COTTERILL, J. H. 'On the equilibrium of arched ribs of uniform section', *Phil. Mag.*, 4th ser., **29**, 1865, 380, 430

[1] See ref. 24.

8 JENKIN, F. 'Braced arches and suspension bridges', *Proc. R. Scot. Soc. Arts*, 1869

9 MOHR, O. 'Beitrag zur Theorie des Fachwerks', *Z. d. Arch. u. Ing. Ver. Z. Hannover*, 1874

10 LEVY, M. *La Statique Graphique*, Gauthier-Villars, Paris 1874

11 CASTIGLIANO, A. 'Nuova teoria intorno all'equilibro dei systemi elastici', *Trans. Acad. Sc. Turin*, **10**, 1875, 380

12 COTTERILL, J. H. *Applied Mechanics*, 1st ed., Macmillan, London 1884

13 MÜLLER-BRESLAU, H. F. B. *Die Neueren Methoden der Festigkeitslehre und der Statik der Baukonstruktionen*, Kroner, Leipzig 1886 (matrices are used)

14 CROTTI, F. *La Teoria dell Elasticita*, Hoepli, Milan 1888

15 ENGESSER, F. 'Ueber Statisch Unbestimmte Trager bei Beliebigem Formanderungs-Gesetze und uber der Satz von der Kleinsten Erganzungsarbeit', *Z. d. Arch. u. Ing. Ver. Z. Hannover*, **35**, 1889, col. 733

16 TODHUNTER, I. and PEARSON, K. *A History of the Theory of Elasticity and of the Strength of Materials*, Vol. I, 1886, and Vol. II, 1893, Cambridge University Press

17 EWING, J. A. *The Strength of Materials*, 1st ed., Cambridge University Press 1899

18 OSTENFELD, A. 'The deformation method of treating statically indeterminate structures', *Der Eisenbau*, **12**, 1921, 275 (*I.C.E. Engineering Abstracts*, **11**, 1922, 69)

19 GRIMM, C. R. *Secondary Stresses in Bridge Trusses*, Wiley, New York 1908

20 OSTENFELD, A. *Die Deformationsmethode*, Springer, Berlin 1926

21 BEGGS, G. E. 'An accurate mechanical solution of statically-indeterminate structures by use of paper models and special gages', *Proc. Amer. Conc. Inst.*, **18**, 1922, 58

22 WESTERGAARD, H. M. 'One hundred and fifty years advance in structural analysis', *Trans. Am. Soc. C.E.*, **94**, 1930, 226

23 SOUTHWELL, R. V. *Relaxation Methods in Engineering Science*, Oxford University Press 1940

24 LIVESEY, R. K. 'Analysis of rigid frames by an electronic digital computer', *Engineering*, **176**, Aug. 1953, 230 and 277

25 TIMOSHENKO, S. *History of Strength of Materials*, McGraw-Hill, New York 1953

26 CHARLTON, T. M. 'Statically-indeterminate frames: the two basic approaches to analysis', *Engineering*, **182**, 1956, 822–3

27 CHARLTON, T. M. *Energy Principles in Applied Statics*, Blackie, London 1959

28 CHARLTON, T. M. *Analysis of Statically-indeterminate Frameworks*, Longmans 1961 and Wiley, New York 1962

29 CHARLTON, T. M. 'Some early work on energy methods in theory of structures', *Nature*, **196**, 1962, 734–6

Appendix II[1]

Maxwell–Michell theory of minimum weight of structures

In 1869[2] Maxwell stated the theorem:

> In any system of points in equilibrium in a plane under the action of repulsions and attractions, the sum of the products of each attraction, multiplied by the distance of the points between which it acts, is equal to the sum of the products of the repulsions multiplied each by the distance of the points between which it acts.

He applied his theorem to plane pin-jointed frameworks as follows:

> In a plane frame, loaded with weights in any manner, and supported by vertical thrusts, each weight must be regarded as attracted towards a horizontal base line, and each support of the frame as repelled from that line. Hence the following rule:
>
> Multiply each load by the height of the point at which it acts, and each tension by the length of the piece on which it acts, and add all these products together.
>
> Then multiply the vertical pressure on the supports of the frame each by the height at which it acts, and each pressure by the length of the piece on which it acts, and add the products together. This sum will be equal to the former sum. If the thrusts which support the frame are not vertical, the horizontal components must be treated as tensions or pressures borne by the foundations of the structure, or by the Earth itself.
>
> The importance of this theorem to the engineer arises from the circumstance that the strength of a piece is in general proportional to its section, so that if the strength of each piece is proportional to the stress which it has to bear, its weight will be proportional to the product of the stress multiplied by the length of the piece. Hence these sums of products give an estimate of

---

[1] Based on an article by the author, *Nature*, **200**, 1963, 251–2, and included here by kind permission of the editors of the journal.
[2] *Scientific Papers*, vol. 2, p. 175 (1869).

the total quantity of material which must be used in sustaining tension and pressure respectively.

Maxwell proved the theorem applied to frames by an argument using the concept of virtual work.

In 1904[1] Michell deduced from Maxwell's theorem that:

A frame therefore attains the limit of economy of material possible in any frame structure under the same applied forces, if the space occupied by it can be subjected to an appropriate small deformation, such that the strains in all the bars of the frame are increased by equal fractions of their lengths, not less than the fractional change of length of any element of the space.

Now, the virtual work equations of all the joints of a framework may be added to give the virtual work equation of the whole, as follows[2]:

$$-\sum fel + \sum Fed = 0 \tag{1}$$

in which the summations include all bars and all external forces F, respectively. Here the virtual work of the bars is expressed in terms of their forces f and virtual changes in lengths, el, where e is a small fraction and the minus sign in equation (1) is to denote work done on a bar when it is subjected to a virtual change in length of the same sign as the force which it exerts, for example, virtual extension of a bar in tension. The virtual displacements of the external forces F (loads and reactions), ed, are in the form of a fraction e times a length d which specifies the point of application of the respective force.

Consideration of equation (1) indicates that if the system of virtual displacements of a framework is such that the loaded framework is imagined to shrink uniformly so that the same fraction e is applicable throughout, then:

$$\sum fl = \sum Fd \tag{2}$$

This is in accordance with Maxwell's concept for the purpose of proving his theorem applied to frameworks, as set out in the first paragraph.

It is clear that for a definite system of loads and reactions the points of application of which are defined, the quantity on the right-hand side of equation (2) is constant, say, K. That is, for all possible frameworks for a given function (i.e. same loading and points of support) the difference between the sum of the products of the forces in the tension members and their lengths and that of the products of the forces in the compression members and their lengths is constant. This is an important principle in relation to the least-weight aspect.

Michell writes equation (2) as follows:

$$\sum f_p l_p - \sum f_q l_q = K(= \sum fl) \tag{3}$$

---

[1] *Phil. Mag.*, 6th Ser., 1904, p. 589.
[2] The symbols used herein are those used by Michell.

where now f_p is the numerical value of the tension in any tie bar of length l_p and f_q is the numerical value of the thrust in any strut of length l_q. He then goes on to say that if P is the greatest allowable tensile stress and Q is the greatest allowable compressive stress, the least volume of material in any frame (among those suitable for withstanding the specified system of loads and reactions) consistent with security is:

$$V = \sum l_p \frac{f_p}{P} + \sum l_q \frac{f_q}{Q} \tag{4}$$

The condition for V to be a minimum is the same as that for the expression:

$$2PQV + (P - Q)K \tag{5}$$

to be a minimum because V is the only variable. Substitution from equations (3) and (4) in (5) shows, however, that V is a minimum when:

$$(P + Q)(\sum f_p l_p + \sum f_q l_q) \tag{6}$$

is least. That is, V is least when the arithmetical sum of the products of lengths of bars and their forces is least. This means that of all the frames which could be constructed in the space bounded by the points of application of the loads and the reactions, that which has the least volume of material has the smallest arithmetical sum of bar force and length products. (This condition follows directly from equation (4) if $P = Q$.)

Michell proceeds to derive the condition for $\sum f_p l_p + \sum f_q l_q$ to be least. His method is rather tortuous and abstract and it is replaced here by what is believed to be a more practical, if less general, approach.

Confining attention to the problem of the framework with the least volume of material to support a specified system of loads acting at specified points, with the aid of a specific system of reactions, the 'space' available is defined by the points of application of the loads and reactions. Now the virtual work of the loads and reactions due to specified virtual displacements of their points of application causing virtual displacements throughout the structural space such that no bar suffers a change of strain numerically greater than ϵ is a definite quantity W. The virtual work of the bars of any framework in the space for which the virtual displacements are compatible with those specified for the loads and reactions must then be equal to W, in spite of the fact that the number of bars and geometry of each possible frame are different. Thus:

$$W = \sum efl \tag{7}$$

(where the summation is, of course, carried out algebraically taking account of the signs of e and f and includes all bars) for each of all possible frames within the limits of the space.

Having regard to equation (7) it appears that the frameworks for which each of the terms $efl\,(e \leqslant \epsilon)$ has the same sign, that is, ties subjected to virtual extensions, struts to virtual contractions, make up the range of frame-works with the smaller values of $\sum f_p l_p + \sum f_q l_q = \sum |f| \, l$. If, in addition, all the virtual strains of the bars can be increases of the same amount ϵ, it follows that $\sum f_p l_p + \sum f_q l_q = W/\epsilon$ is the smallest possible value. This means that a framework attains the limit of economy of material possible in any framework subjected to the same system of applied forces with specified points of applica-tion, if the coordinates of its joints with reference to any specified origin can be imagined subjected to small changes (consistent with uniform small deformation of the space occupied by the structure) such that the strains of all the bars receive equal virtual increases. (It is emphasised that strain is defined as a fraction: an increase of strain implies additional extension of contraction, depending on whether the bar functions as a tie or a strut.) Although the analysis is presented with reference to plane pin-jointed frameworks the implications appear to be quite general and it seems to follow from equation (2) that the members of economical frameworks will be wholly in tension or wholly in compression.

Practical application of the theory of minimum weight structures in general is in its infancy. For further details of the subject the reader is referred to the works of Cox and Hemp which are listed below. It would appear, however, that considerations of self-weight and erection requirements seriously limit its potentiality for civil engineering structures in general.

REFERENCES

COX, H. L. *The Design of Structures of Least Weight*, Pergamon, London 1965
HEMP, W. S. 'Studies in the theory of Michell structures', *Proc. XIth Int. Cong. App. Mech.*, Munich 1964

Bibliography

1 MATHESON, J. A. L. *Hyperstatic Structures*, Butterworth 1959
2 CASE, J. and CHILVER, A. H. *Strength of Materials*, Arnold 1959
3 THADANI, B. N. *Structural Mechanics*, Asia 1964
4 GREGORY, M. S. *Linear Framed Structures*, Longmans 1966
5 SOUTHWELL, R. V. *Theory of Elasticity*, Oxford University Press 1941
6 CROSS, H. *Arches, Continuous Frames, Columns and Conduits*, Illinois 1963
7 SPIELVOGEL, S. W. *Piping Stress Calculations Simplified*, McGraw-Hill 1943
8 DUGAS, R. *A History of Mechanics*, Routledge and Kegan Paul 1955
9 GREGORY, M. S. *Elastic Instability*, Spon 1967
10 BROWN, E. H. *Structural Analysis*, Vol. I, Longmans 1967
11 LIVESLEY, R. K. *Matrix Methods of Structural Analysis*, Pergamon 1964
12 NEAL, B. G. *Structural Theorems and their Applications*, Pergamon 1964
13 CHARLTON, T. M. *Model Analysis of Plane Structures*, Pergamon 1966
14 RAYLEIGH, Baron. *Theory of Sound*, Vols. 1 and 2, 2nd ed., Macmillan 1894
15 ALLEN, D. N. DE G. *Relaxation Methods*, McGraw-Hill 1951
16 FRAZER, R. A., DUNCAN, W. J. and COLLAR, A. R. *Elementary Matrices*, Cambridge University Press 1938
17 VON KARMAN, T. *Mathematical Methods in Engineering*, McGraw-Hill 1940
18 YLINEN, A. *Kimmo-Ja Lujuusoppi*, Helsinki 1965
19 CHARLTON, T. M. *Energy Principles in Applied Statics*, Blackie 1959
20 LIVESLEY, R. K. and CHARLTON, T. M. 'The use of a digital computer with particular reference to the analysis of structures', *Trans. N.E.C. Inst.*, **71**, 1954, 67
21 MORRIS, J. *Proc. I. Mech. E.*, **155**, 1964, 38
22 PUGSLEY, A. G. *The Theory of Suspension Bridges*, Arnold 1957 (2nd ed. 1968)
23 BOWEN, C. F. P. and CHARLTON, T. M. 'A note on the approximate analysis of suspension bridges', *The Structural Engineer*, **45**, 1967, 241

Index